SUPPLEMENTAL MATERIAL FOR *THE JOY OF CHEMISTRY*

Additional information for
The Amazing Science of Familiar Things

by Daniel J. Berger

Printed using CreateSpace
A division of Amazon.com

Copyright © 2013 Daniel J. Berger
All rights reserved.
ISBN: 1492989975
ISBN-13: 978-1492989974

For Peter and James, with thanks.
If I am a chemist, it is because of them.

Table of Contents

Preface for students 1

Part 1, Chapter 1: "Electrons and Atoms, Elephants and Fleas" 3
    Scientific models 3
    The problem of measurement 5
    "Atoms are empty" – NOT! 5
    Atomic structure 6
    Vocabulary 7
    Questions 8

Part 1, Chapter 2: "Periodically Speaking" 9
    Atomic mass and the atomic nucleus 9
    Electrons and valence 12
    Vocabulary 15
    Exercises 16

Part 1, Chapter 3: "Reason, reactions and redox" 17
    Rules for assigning oxidation state 18
    Vocabulary 19
    Exercises 20

Part 1, Chapter 4: "The basic stuff" 21
    Vocabulary 24
    Questions and exercises 25

Part 1, Chapter 5: "Chemical partners: who does what to whom" 27
    Molecular polarity 29
    Hard and soft water 30
    Vocabulary 32
    Questions 33

Part 1, Chapter 6: "The tie that binds, the chemicals that bond" 34
    Types of bonds 34
    Using the periodic table to predict bonding 36
    Electrical conduction 39
    Vocabulary 41
    Questions 42

Part 1, Chapter 7: "Sticking to principles" 43
    The law of definite proportions 43
    Condensed formulas 44
    Isomers 45
    Conservation of mass 46
    Vocabulary 46
    Questions 47

Part 1, Chapter 8: "Slipping and sliding, intermolecularly" 49
    More about London (dispersion) forces 50
    Vocabulary 51
    Questions 52

Part 1, Chapter 9: "Concentration—on being alone together" 53
    The mole as a chemical unit 53
    Concentration 54
    Vocabulary 56
    Exercises 57

Part 1, Chapter 10: "It's a gas" 59
    The Kelvin temperature scale 60
    How to solve proportionality questions 62
    Vocabulary 63
    Questions 64

Part 1, Chapter 11: "When gases put on airs" 65
    Free radicals 65
    Catalysts in chemistry 67
    For example: photosynthesis and respiration 68
    Vocabulary 69
    Questions 70

Part 1, Chapter 12: "Crystal clear chemistry" 71
    Molecules in 3-D 71
    Crystal structures 74
    Adhesives 74
    Vocabulary 75
    Questions 76

Part 1, Chapter 13: "When matter heats up" 77
    Enthalpy 77
    Entropy 80
    Free Energy 81
    Vocabulary 83
    Questions 84

Part 1, Chapter 14: "A whole new phase" 85
    Phase changes as reversible processes 86
    Phase diagrams 86
    Vocabulary 88
    Questions 88

Part 1, Chapter 15: "Equilibrium — chemistry's two-way street" 91
    Dynamic equilibrium 91
    Equilibrium mixtures 91
    Unbalanced equilibrium 92
    Reversible and irreversible reactions 93
    Energy, entropy and equilibrium 94
    Vocabulary 95
    Questions 95

Part 1, Chapter 16: "Colligative properties — strength in numbers" 97
    Osmotic pressure 97
    Reverse osmosis 98
    Vocabulary 99
    Questions 100

Part 1, Chapter 17: "Chemical kinetics — a veritable explosion" ... 101
  Activation energy ... 103
  Vocabulary ... 104
  Questions ... 105

Part 1, Chapter 18: "Electrons and photons — turning on the light" ... 107
  How photochemistry works ... 107
  How electrochemistry works ... 108
  Vocabulary ... 113
  Questions ... 114

Part 2, Chapter 1: "Simply organic" 115
    Organic compounds are modular 116
    Organic molecules can be chiral 117
    Vocabulary 119

Section 2, Chapter 2: "Chemistry rocks" 121
    Element groupings 122
    Transition metals and coordination chemistry 123
    The main-group elements ("representative elements") 123
    Radioactive elements and "radiochemistry" 124
    Vocabulary 128

Section 2, Chapter 3: "The body of chemistry meets the chemistry of the body" 129
    Polymers 129
    Biopolymers 130
    Vocabulary 138

Section 2, Chapter 4: "Chemist as analyst" 139
    Precision vs. accuracy 139
    Analytical techniques: separation and identification 140
    Vocabulary 142

Answers to selected problems and questions 143

    Chapter 1 143

    Chapter 2 144

    Chapter 3 145

    Chapter 4 145

    Chapter 5 146

    Chapter 6 147

    Chapter 7 148

    Chapter 8 149

    Chapter 9 150

    Chapter 10 151

    Chapter 11 152

    Chapter 12 153

    Chapter 13 154

    Chapter 14 155

    Chapter 15 156

    Chapter 16 157

    Chapter 17 158

    Chapter 18 158

Preface for Students

This is a supplement to Cathy Cobb and Monty Fetterolf's *The Joy of Chemistry*, the textbook for NSC 105, The Chemistry of Everything, at Bluffton University.

The Joy of Chemistry is usually referred to as "Cobb & Fetterolf" in this text. **You should read each chapter in Cobb & Fetterolf *before* reading the corresponding chapter in this supplement.** Each chapter in the text has a corresponding chapter in this supplement, which provides additional discussion of important concepts, a vocabulary list, and questions or exercises. Please note that some of the supplemental material is truly *supplemental*, that is, it provides information that is *not* in Cobb & Fetterolf.

On the other hand, you are also responsible for everything in Cobb & Fetterolf, and *most of that material is not reproduced* in these supplements. Some chapters of this supplement are quite short, as there is little to add to the corresponding chapter in Cobb & Fetterolf.

Comments are welcome (and worth bonus points)

Your comments are solicited for each chapter of this supplement, and are worth up to 3 bonus quiz points per chapter. The number of such points awarded will be based on how helpful your comments are to me in revising and improving this material.

Comments should pertain to

- Readability. How easy is it to read and comprehend the supplement?
- Usefulness. Does the supplement help you understand what's being stated in *The Joy of Chemistry*?
- Suggestions for expansion, or for useful questions, problems or vocabulary words. Note that these will usually not be placed into the supplements, but will appear on the course's Moodle page.

A note about *The Joy of Chemistry*

I realize that the text has shortcomings; if it didn't, this supplement wouldn't be necessary.

This text was chosen partly for being accessible to the general reader with no science background, partly because of the many home chemistry demonstrations it contains, and partly for its price. I imagined that students would appreciate a book that costs only 10-20% of a more traditional textbook.

For students interested in a more AP-chemistry-like experience, I recommend Robert Bruce Thompson's *Illustrated Guide to Home Chemistry Experiments* (O'Reilly, 2008). Or better yet, take a majors-level general chemistry course.

D. B.
November 1, 2013

PART 1, CHAPTER 1: "ELECTRONS AND ATOMS, ELEPHANTS AND FLEAS"

Chapter supplements such as this are not intended to replace reading the corresponding chapter in *The Joy of Chemistry*! Instead, they are intended to provide additional information, or to correct oversimplifications that I think it important to get in more detail than Cobb and Fetterolf provide.

In Part 1, Chapter 1, two things I think worth discussing in more depth are scientific models and the structure of atoms.

Scientific models

Whenever we discuss anything in the world around us, we do it with symbols: words, for example. The word **d-o-g** is not a dog, but a symbol that represents things that real, individual dogs have in common.

If we want to make the description more specific, we add more words to specify "a male basset" or "that brown and white dog on the floor over there." Yet the descriptions themselves are symbolic, ways of representing or signifying the things that they describe; they are valuable only in so far as they help us to think about the world. *Hund*, *pies* or собака signify the same concept as "dog," but only if you speak German, Polish or Russian; to an English speaker, they may confuse rather than communicate.

Belgian artist René Magritte illustrated this in his painting, "The Treachery of Images" ("This is not a pipe"), shown at right. The point is that a picture of a pipe is not the same as a pipe, but a *symbol* that *represents* a pipe. As Magritte himself noted, "Could you stuff my pipe? ... if I had written on my picture 'This is a pipe' I'd have been lying!"

Ceci n'est pas une pipe.
Source: *Wikimedia*

In science, we use symbols to talk about the world; we also use models, which are symbolic representations of things we observe, or of common features of things we observe. Symbols and models are not the things being represented, but can be thought of as metaphors or placeholders for those things. Some examples of models and symbols that you will encounter in this course are

- Atomic symbols such as Fe or O or Na.
- Concepts such as "chemical element" or "gas" or "molecule."
- Three-dimensional models of molecular structure – molecules do have real structures, but those structures are represented using models of various sorts.

Scientific models operate at different levels, from simple visual models that you can draw on paper or hold in your hand, to complex mathematical models that can only be solved using computers. Normally we use the simplest model that will adequately explain what we want to explain,* with the caveat that the model must make useful and testable predictions.

Chemists in particular, but also other scientists, may use two or more different models for the same thing, depending on the particular properties that they want to think about. A successful model usually does not try to represent everything about what is being studied, but only the parts that are considered interesting. Models can be successful without being exact, and more exact models can sometimes be unsuccessful because they are too complex to be helpful.

---

* This is called Occam's razor, or the principle of parsimony. It is often, mistakenly, thought that Occam's razor says the simplest possible explanation *must* be true. Instead, William of Occam was warning against constructing models that are so complicated that we can't understand them, as well as setting out an aesthetic principle: "simpler explanations are more elegant."

The problem of measurement

Scientific instruments are never 100% certain, just as it's impossible to get anything 100% pure. You pay high prices for almost anything that's purer than about 99%; "pure" (24 carat) gold is at least 99.9% gold.

Source: Wikimedia

All instruments have detection limits, values below which they cannot detect anything. Thus, if an instrument can't find a pollutant, that doesn't mean that the pollutant is not there at all; it simply sets an upper bound on what concentration that pollutant could have. If you can't detect dioxin in an environmental sample, and the detection limit is 1 part per billion, then all you can say is that the dioxin level is less than 1 part per billion — **not** that there is no dioxin at all.

Impurities can be supremely important. A gripping fictional example is Robert Louis Stevenson's *The Strange Case of Dr. Jekyll and Mr. Hyde*, which turns on trace impurities in a supposedly pure substance. Quality control is better these days...

"Atoms are empty" – NOT!

Cobb and Fetterolf assert that, because nuclei and electrons are both so small in relation to the full size of atoms, that atoms are mostly empty space. But this is true only if you use the particle model to think about electrons.

In the particle model, we think about electrons and nuclei as small, discrete objects of a particular size. But quantum mechanics tells us that *every* physical object has an associated waveform that, among other things, describes its effective volume.* The size of this waveform is inversely proportional to the object's mass: more massive objects have smaller waveforms.

---

* This is often called "wave-particle duality," a model that says that electrons are *both* particles and waves; but this, too, is metaphorical. Electrons have some properties that we associate with visible objects, like baseballs, and some that we associate with visible waves, like water waves, but they are not "both particles and waves." They are electrons.

For macroscopic objects such as sand grains or baseballs, the waveform does not add to the object's size at all – the wavelength associated with a baseball is a billion trillion trillion times *smaller* than the baseball itself. Atomic nuclei have associated waveforms that are similar to their own size, so that again, we don't usually need to consider the wavelength of a nucleus.* We can consider an atomic nucleus to be a conventional object, like a baseball (though much, much smaller).

But the wavelength associated with an electron is a trillion times *larger* than the size of the electron itself; it is about the same size as a whole atom! This means that atoms are *not* empty; they are filled with "electron density." The reason that your hand doesn't sink into a tabletop is because the electrons in your hand and the electrons in the tabletop fill their respective atoms, and will not interpenetrate.† Electron density will be further discussed in *The Joy of Chemistry*, Part 1, Chapters 2 and 6.

Further information about electrons
Why we can say an electron is a wave, not a particle:
hyperphysics.phy-astr.gsu.edu/hbase/debrog.html
How electron waves behave in atoms:
hyperphysics.phy-astr.gsu.edu/hbase/debrog.html

Atomic structure

The discussion of atomic structure in *The Joy of Chemistry* is adequate; here is a summation, with some additional information:

The atomic nucleus contains **protons** (massive; positive electrical charge) and **neutrons** (massive; no electrical charge) are tightly bound to each other. These particles are known as **nucleons** because they are found in the nucleus. The number of protons in an atom defines its chemical identity. This is not because protons participate in chemical

---

* For very small nuclei, such as hydrogen, this is not quite true; there are some properties of hydrogen atoms that can only be explained using a waveform for their atomic nuclei. But these properties are beyond the scope of this course.
† The reason for this is *not* electrostatic repulsion between electrons, but a quantum mechanical property called the Pauli Exclusion Principle that keeps quantum-mechanically similar electrons from occupying the same space.

reactions, but because the number of protons controls the number of electrons.

Electrons (at least 3000 times lighter than protons and neutrons; negative electrical charge) occupy the space outside the nucleus. Atoms must have the same number of electrons as they have protons, in order to be electrically neutral. Electron exchange and sharing is what lies at the root of chemical interactions, as we will see in subsequent chapters.

Ions are atoms (or molecules, clusters of atoms) that have more or fewer electrons than their number of protons, so that they contain unbalanced electrical charge. Ions come in two varieties:
- **Cations** are positively-charged ions; they have more protons than electrons.
- **Anions** are negatively-charged ions; they have more electrons than protons.

Vocabulary

You should write your own definitions for these words, based on the textbook and this outline:

Scientific model

Atom

Atomic nucleus

Proton

Neutron

Nucleon

Electron

Ion

Cation

Anion

Questions

You will need to read both this supplement and the Introduction and Chapter 1 of Cobb & Fetterolf in order to answer these questions.

1. What constitutes a chemical change? How is it different from a physical change? How would you tell?

2. If something with one name changes into something with another name, is that always a chemical change? Can you think of an example when it is not?

3. If a substance is not detected, does that mean it is not there? Why would we say, for example, that "there are no pesticides in this organic produce"? Is that a reasonable thing to say?

4. In what way is science simply an extension of the sort of thinking we do about things every day? Can you think of an example of a time when you thought through something carefully?

PART 1, CHAPTER 2: "PERIODICALLY SPEAKING"

Different chemical substances behave differently. This behavior can be understood and organized in a useful way.

Atomic mass and the atomic nucleus

Mass is the amount of matter (or the degree to which matter resists force); weight is the force exerted by gravity on a mass. The key technique separating modern chemistry from alchemy is keeping track of mass as one thing is changed into another. Doing this allowed chemists to begin identifying elements with less guesswork.

Each chemical element has a "characteristic mass" (now called its "atomic mass") that can be followed through all its reactions. This was known by about 1800. When Dmitri Mendeleev (ca. 1860) sorted the elements by their characteristic masses, he found that chemical properties repeat in a periodic fashion. This led to the first Periodic Table, published in 1869. I repeat: **the periodic table has the structure that it has because of the fact that, when elements are sorted in order of increasing characteristic mass, chemical properties repeat in a periodic fashion.**

There are a few anomalies: the characteristic mass of tellurium is greater than that of iodine; the characteristic mass of cobalt is slightly greater than that of nickel. Mendeleev, boldly following the chemical properties, said that this meant that the measurements of the atomic masses of those elements were incorrect. But he was wrong.

The true basis of the periodic table was discovered in about 1913 by Henry Moseley: **atomic number**. Atomic number is the number of protons in the nucleus, and therefore the number of electrons in a neutral atom. Every element has its characteristic atomic number, Z: for example, hydrogen has Z=1, calcium has Z=20, tellurium has Z=52, and iodine has Z=53. Atomic number was found to correspond directly to an atom's ordering in the periodic table.

But the discovery of atomic number left some holes in our knowledge of atomic structure. The first is this: why does hydrogen (Z=1) have a characteristic mass of one, but helium (Z=2) has a characteristic mass of **four** and lithium (Z=3) has a characteristic mass of **seven**? To explain this, an electrically neutral particle, the neutron, was postulated, with a mass equal to that of the proton; the neutron was first detected in 1930-1931.*

Secondly, why aren't atomic masses exact integers, since 1 atomic mass unit seems to be equal to the mass of one proton or one neutron? This problem has been known since the 19th Century, but was at first ascribed to experimental imprecision. However, more precise measurements simply made the problem more acute until the advent of mass spectrometry in about 1913, allowing the masses of individual atoms to be measured. When this was first done, it was discovered that atoms of the same element do not all have exactly the same mass. This phenomenon was not due to impurities in the element, since impurities showed up at their own characteristic masses. Instead, what had been discovered was the existence of **isotopes**: different atoms of the same element (having the same atomic number) that have different masses.

---

* Until its discovery, the neutron was not the consensus theory in atomic physics. Instead, it was thought that an atom's mass came from having the same number of protons, so that (for example) helium had four protons in the nucleus. Excess protons were thought to be balanced by "nuclear electrons," so that in a helium nucleus, two electrons plus four protons gave a nuclear charge of +2. These negatively-charged electrons would also serve to keep the positively-charge protons from flying apart. This model, while incorrect, is interestingly confirmed by the fact that *free* neutrons (that is, neutrons that are not inside atomic nuclei) are unstable and quickly decay into protons and electrons, with a lifetime of about 15 minutes. This is also the source of some radioactivity, in atomic nuclei with more neutrons than they need: see the supplemental notes for Part 2, Chapter 2.

Isotopes are NOT due to ionization. Electrons mass almost nothing – 3000 times less than a proton. All the electrons in an atom put together add up to only a small fraction of the mass of a single proton in its nucleus.

Isotopes ARE due to the number of neutrons in the nucleus. Atomic nuclei can have different numbers of neutrons even if they have the same number of protons, because neutrons do not have an electrical charge. As we will see later, it is the electrical charge of an atom's nucleus, rather than its mass, that dictates its chemistry.

Since neutrons weigh the same as protons, different numbers of neutrons show up as a difference in atomic mass. The number of protons plus the number of neutrons equals the **mass number** of an isotope, which is then described using the element's symbol and the mass number. For example, hydrogen has two stable isotopes: ^{1}H or hydrogen-1 (one proton, zero neutrons) and ^{2}H (one proton, one neutron). Lithium also has two stable isotopes: ^{6}Li (three protons, three neutrons) and ^{7}Li (three protons, four neutrons).

The characteristic mass of an element is a weighted average of isotopic masses. Characteristic masses of elements are (or were) measured in bulk: zillions of atoms at a time. So what we observe as the characteristic mass is an **average** of all the isotope masses in the sample of an element.

For example, hydrogen is mostly ^{1}H, with small amounts of ^{2}H; so the characteristic mass (hereafter, *atomic mass*) of hydrogen is a little bit more than 1. Meanwhile, lithium is mostly ^{7}Li, with a fair amount of ^{6}Li, so the atomic mass of lithium is a little less than 7.

So why are there neutrons in the nucleus at all? The strong nuclear force (or "strong force", which acts only within atomic nuclei) is what keeps all the positively-charged protons in any given atomic nucleus from flying violently apart, due to their electrostatic repulsion for each other. Neutrons, which attract other nucleons via the strong force but don't have any electrical charge, increase the amount of strong force holding protons in the nucleus, without contributing any repulsive electrostatic force.

It so happens that two protons don't generate enough strong force to keep themselves from flying apart, so that helium-2 is not a stable isotope. But add a single neutron, and the protons are able to hang together: helium-3 *is* stable. The most stable isotope of helium is helium-4, with two protons and two neutrons.

As you pack more and more protons into the nucleus, you need to pack in more neutrons as well. After about 20 protons, it's no longer enough to have a roughly equal number of neutrons (as in carbon-12, fluorine-19, sulfur-32 or calcium-40). You need more neutrons than protons to keep things together: iron-56 (26 protons, 30 neutrons), tin-114 (50 protons, 64 neutrons), lead-208 (82 protons, 126 neutrons), uranium-238 (92 protons, 146 neutrons).

> Nuclear binding energy, the energy that holds protons and neutrons together, contributes a small amount of mass. This is due to the fact that mass is a very concentrated form of energy ($E = mc^2$). Nuclei that are more energetically stable (and thus require less binding energy) have a smaller average mass per nucleon, and so their exact mass is likely to be a bit less than their mass number. Nuclei that are less stable require more binding energy and will have a higher exact mass, typically a bit more than their mass number. This variation is part of the reason that observed atomic masses are not integers, even if they are measured for single atoms.

While chemists recognize that an element symbol or element name describes its atomic number, since each element has its own atomic number, physicists like to include the atomic number in atomic symbols as a subscript, so that (for example) hydrogen-2 or ²H is represented as ^2_1H, while oxygen-16 or ¹⁶O is represented as $^{16}_8O$.

Electrons and valence

Valence electrons are the outer electrons of an atom; they are the electrons that participate in chemistry. Since electrons within atoms are distributed in "shells" that can each hold only so many electrons and no more – and only the outer shell's electrons participate in chemistry – the chemistry of each element is governed by its atomic number, because the atomic number dictates the number of electrons in a neutral atom and, indirectly, the number of valence electrons.

Valence is the number of "bonds" (see Chapter 6) that an atom can form in compounds. The International Union of Pure and Applied Chemistry defines valence as "The maximum number of univalent atoms (such as hydrogen or chlorine atoms) that may combine with an atom of the element under consideration." Thus, the concept of valence is closely tied to the formation of electron-pair bonds by an atom (see Chapter 6). Valence is normally expressed as a positive integer, and different elements have different normal valences: for example, carbon always has a valence of 4, while hydrogen or fluorine always has a valence of 1.

The Periodic Table is divided into columns called *families*; each element within a family has the same number of **valence electrons**, and therefore (usually) the same valence, as any other element within the same family. For example, all elements in Family 1 (the "alkali metals") have one valence electron. All elements in Family 17 (the "halogens") have seven valence electrons. All elements in Family 18 (the "noble gases") have eight valence electrons.*

Family numbers are not necessarily equal to the number of valence electrons. This is because the International Union of Pure and Applied Chemistry decided that the old system of numbering families – based upon the most common valence, or the number of valence electrons, within those families – was too confusing. **However,** in family numbers greater than 10, you can get the true number of valence electrons by subtracting ten from the family number.

The **valence** is not equal to the number of valence electrons, but often follows the *octet rule*: how many electrons you need to add or subtract to reach the nearest octet—that is, the same number of valence electrons as the nearest noble gas (Group 18) element. For example, alkali metals (Family 1) have valences of 1 because *removing* one electron gives the nearest octet. Chalcogens (Family 16) have most common valences of 2 because *adding* two electrons gives the nearest octet. And noble gases (Family 18) have valences of zero *because they ARE the nearest octet*. See the discussion of the *shell model* and the *octet rule* in the text.

The text elides a peculiarity in how shells are filled. Each shell contains the same number of subshells as its shell number; that is, shell #1 contains one subshell, #2 contains two subshells, and so forth. Subshells each hold four more electrons than the previous one, so that the first subshell of each shell holds two electrons, the second subshell holds six, the third ten, and so on.

---

* Hydrogen and helium are special cases. Hydrogen has one valence electron, and helium has two, but they are grouped where they are – hydrogen sometimes in Group 1 and sometimes in Group 17, and helium always in Group 18 – because of their chemical properties.

The first two subshells of each shell are filled within the corresponding row, for a total of eight (two plus six) electrons; this is why the noble gases other than helium have eight valence electrons, the "octet." But

- The third subshell of the third shell is not filled until the fourth row.
- The third subshell of the fourth shell is not filled until the fifth row, and the fourth subshell of the fourth shell is not filled until the sixth row.
- The third subshell of the fifth shell is not filled until the sixth row, the fourth subshell of the fifth shell is not filled until the seventh row, and the fifth subshell of the fifth shell will not be filled until we find elements in the eighth row.

The fact that, for example, argon has 18 electrons rather than 28 (as it would if it had a completely full third shell) can be a little confusing... but, under carefully controlled experimental conditions, nature does whatever she wants. Chemistry is an empirical science.

Remember that the concept of valence is a way of describing what we find in the natural world, not necessarily a hard-and-fast rule. This can be seen by the fact that many elements have more than one observed valence. The transition metals are often so; iron (Fe) has common valences of 2 and 3, while vanadium can have valences of 2, 3, 4 or 5. Non-metals can also have multiple valences; for example, phosphorus can be 3 or 5, and sulfur can have valences of 2, 4 or 6.

Vocabulary

You should write your own definitions for these words, based on the textbook and this outline:

Atomic mass

Atomic number

Characteristic mass of an element

Family (as a structural feature of the periodic table)

Isotope

Mass number

Metal

Metalloid

Non-metal

Octet rule

Period (as a structural feature of the periodic table)

Shell (as in the shell model of atomic structure)

Valence

Valence electrons

Valence shell

Exercises

1. How many <u>valence electrons</u> does each of the following elements have? The atomic number and atomic symbol are given for each.

   a. Sodium (Na, 11)

   b. Chlorine (Cl, 17)

   c. Hydrogen (H, 1)

   d. Oxygen (O, 8)

   e. Nitrogen (N, 7)

   f. Scandium (Sc, 21)

   g. Fluorine (F, 9)

   h. Sulfur (S, 16)

2. Using the octet rule, predict the <u>common valence</u> of each element listed below, that is, the number of chemical bonds it is likely to form. The atomic symbol and atomic number are given for each.

   a. Sodium (Na, 11)

   b. Chlorine (Cl, 17)

   c. Hydrogen (H, 1)

   d. Oxygen (O, 8)

   e. Nitrogen (N, 7)

   f. Scandium (Sc, 21)

   g. Fluorine (F, 9)

   h. Sulfur (S, 16)

PART 1, CHAPTER 3:
"REASON, REACTIONS AND REDOX"

The text discussion is good. Some additions:

LEO the lion says GER is fine, but seems needlessly complicated to me. An alternate way of remembering oxidation and reduction is OIL RIG:

Oxidation

Is

Loss (of electrons)

Reduction

Is

Gain (of electrons)

Redox is short for "reduction-oxidation." This term reminds us that, in order for something to be oxidized, something else must be reduced. Reactions in which redox takes place are called *redox reactions*.

Oxidation state (also called **oxidation number**) is the formal electrical charge on an atom due to its loss or gain of electrons according to the rules of valence. For example, in iron(II) sulfide, FeS, iron has an *oxidation state* of $+2$, while sulfur has an *oxidation state* of -2.

Oxidation states tend to work according to the octet rule: an element in Family 1 will tend to have an oxidation state of $+1$ (because it needs to lose one negatively-charged electron to attain the nearest octet), while an element in Family 17 will tend to have an oxidation

17

state of -1 (because it needs to gain one negatively-charged electron to attain the nearest octet.)

But there are exceptions. Most of them are covered by the rules below:

Rules for assigning the oxidation state of an atom in a compound of known formula:

First, note that for any electrically-neutral compound, the sum of all the oxidation states of the different atoms must be zero. In CaO, oxygen has an oxidation state of -2 (Rule 2, below) and therefore calcium must have an oxidation state of $+2$.

0. All pure elements, or atoms bound only to other atoms of the same element, have oxidation states of ZERO. **Always.**
1. Fluorine **always** has an oxidation state of -1 when combined with other elements. **Always.**
2. Oxygen **always** has an oxidation state of -2, unless this conflicts with Rules 0 or 1.
   - Example: CaO vs. OF_2.
   - In "peroxides" such as hydrogen peroxide, H_2O_2, oxygen has an oxidation state of -1 because it is bound to itself (H−O−O−H) as well as to some other element. Peroxides are not very stable, and so there are not many compounds of this type.
3. Halogens other than fluorine have oxidation states of -1 unless this conflicts with Rules 0, 1 or 2. (e.g. NaBr vs. BrF or $NaBrO_3$)
4. Hydrogen always has an oxidation state of $+1$ when combined with non-metals and -1 when combined with metals or semimetals. For example, in CH_4 hydrogen is $+1$, while in NaH or BH_3, hydrogen is -1.
5. Non-metals other than hydrogen always have negative oxidation states that follow the octet rule, unless this conflicts with Rules 1 through 4. (For example, consider FeS and SO_2: in FeS, sulfur has a negative oxidation state, while in SO_2 it has a positive oxidation state because of Rule 2.) But see also rule 6, especially when combined with Family 1 or Family 2 metals.

6. Metals almost always have positive oxidation states; these are most easily found by assigning them the charge left over after following Rules 1 through 5. Metals in Families 1 and 2 always follow the octet rule: Family 1 metals are always +1, while Family 2 metals are always +2.

A hard and fast rule: No atom can have an oxidation state that is greater than its number of valence electrons. For example, sulfur – with 6 valence electrons – cannot have an oxidation state greater than +6; silicon cannot be more than +4; scandium cannot be more than +3; sodium cannot be more than +1.

Vocabulary

You should write your own definitions for these words, based on the textbook and this outline:

Oxidation

Reduction

Oxidizing agent

Reducing agent

Redox reaction

Oxidation state / Oxidation number

Exercises

1. What is the oxidation number for tin in stannous fluoride, SnF_2?

2. What is the oxidation number for oxygen in limestone, $CaCO_3$?

3. What is the oxidation number for iron in rust, Fe_2O_3?

4. What is the oxidation number for hydrogen in sulfuric acid, H_2SO_4?   (This   is   more   difficult!)

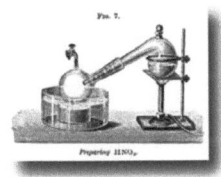

Part 1, Chapter 4: "The basic stuff"

In every acid-base reaction there is an acid and a base. This sounds like a pointless thing to say, but sometimes you don't recognize the acid or the base because it's not something we commonly think of as an acid or base.

For example, when sulfuric acid is dissolved in water, it undergoes an acid-base reaction in which sulfuric acid (H_2SO_4) is the acid and water (H_2O) is the base:

$$H_2SO_4 + 2\,H_2O \rightarrow SO_4^{2-} + 2\,H_3O^+$$

Likewise, when ammonia is dissolved in water, it undergoes an acid-base reaction in which ammonia (NH_3) is the base and water (H_2O) is the acid:

$$NH_3 + H_2O \rightarrow NH_4^+ + OH^-$$

This reaction with water is called *dissociation* and we say that the acid or base **dissociates** when it is dissolved in water. By now, you may have noticed something:

- Acids generate H_3O^+ (*hydronium ion*) in water; they donate H+ to bases.
- Bases generate OH− (hydroxide ion) in water; they accept H+ from acids.
- Every acid-base reaction involves both an acid and a base, just as redox reactions involve both oxidation and reduction.

You also need to be aware of the difference between **strong** and **weak** acids and bases. Strong acids and bases can be dangerous, while weak acids and bases are usually pretty harmless unless they happen to be harmful for other reasons. (For example, hydrogen cyanide is a weak acid — "prussic acid" — but is dangerous because it's poisonous.)

Strong acids and bases generate at least one hydronium or hydroxide ion per molecules, when they are dissolved in water. Sulfuric acid is a strong acid: every single molecule of dissolved sulfuric acid reacts with water to form hydronium ions. Other common strong acids include hydrochloric acid, also called muriatic acid; nitric acid, used to make fertilizers and explosives; and phosphoric acid, used to make fertilizers.

The most common strong bases are compounds containing hydroxide (OH^-) or oxide (O^{2-}) ions. Examples include lye (sodium hydroxide) and lime (calcium oxide). Limewater is a dilute solution of a strong base, calcium hydroxide.

Weak acids and bases generate very little hydronium or hydroxide when they are dissolved in water. Ammonia is a weak base; only about one ten-thousandth of dissolved ammonia molecules reacts with water to form hydroxide ions. Other common weak bases include baking soda (sodium hydrogen carbonate), alkaloid drugs such as codeine and cocaine, and stimulant compounds such as adrenaline, ephedrine, pseudoephedrine, and amphetamines.

Common weak acids are often biologically derived. Examples include not only hydrogen cyanide, found in bitter almonds, but also acetic acid (vinegar) and such pain relievers as aspirin, acetaminophen and ibuprofen.

Indicators are compounds that change color when mixed with an acid or a base; some indicators (such as phenolphthalein, the stuff that is clear in acid but turns pink in base) will only change color with one, either acid or base, while others (such as red cabbage indicator) will change color for both acid and base.

Indicators do this because they are themselves acids or bases. Phenolphthalein is an acid that reacts with bases to form a colored anion. Red cabbage indicator is a purple anthocyanin dye that changes to red when it acts as a base toward an acid, and turns blue-green when it acts as an acid toward a base. Compounds, such as water or anthocyanin dyes, that can react both as an acid toward a base and as a base toward an acid, are called *amphoteric*.

Polyatomic ions follow the rules of valence and oxidation state. It's just that you have to account for the charge on the ion as well. For example,

- Sulfur in sulfur trioxide, SO_3, has an oxidation state of +6: three oxygen atoms, each with an oxidation state of −2 (see Rule 2 in the Chapter 3 notes), make −6, and to balance this, sulfur must be +6.
- Sulfur in sulfate ion, SO_4^{2-}, also has an oxidation state of +6. There are four oxygen atoms, each with an oxidation state of −2, adding up to −8. But the net charge on the sulfate ion is −2, so sulfur only has to balance six of the eight negative charges.

Here's an experiment: get an effervescent antacid tablet, such as Alka-Seltzer™. Why does it fizz when dissolved in water? Why doesn't it fizz in the dry tablet? You may be better able to answer this question after we finish Chapter 5.

Polyatomic ions you should know

All of these ions are found in common household substances.

ion		*corresponds to*	
sulfate	SO_4^{2-}	sulfuric acid	H_2SO_4
nitrate	NO_3^-	nitric acid	HNO_3
carbonate	CO_3^{2-}	carbonic acid	H_2CO_3
bicarbonate	HCO_3^-	carbonic acid	H_2CO_3
phosphate	PO_4^{3-}	phosphoric acid	H_3PO_4
hydroxide	OH^-	water	H_2O
ammonium	NH_4^+	ammonia	NH_3

Vocabulary

You should write your own definitions for these words, based on the textbook and this outline:

Acid / strong and weak acids

Base / strong and weak bases

Indicator

Neutralize

Buffer

Alkali or alkaline

Corrosion

pH scale

Hydronium ion

Hydroxide ion

Polyatomic ion

Questions and exercises

1. What is the difference between a strong acid and a weak acid?

2. When sulfuric acid is dissolved in pure acetic acid (the acid in vinegar), the following reaction takes place:

$$H_2SO_4 + HC_2H_3O_2 \longrightarrow HSO_4^- + H_2C_2H_3O_2^+$$

   sulfuric acid   acetic acid

   In this reaction, which is the acid, and which is the base?

3. In the excerpt from Little Men, the combination of sour cream with baking soda removes the objectionable tastes of both of them. Why does baking soda remove the sour taste from sour cream? Why does sour cream remove the bitter taste from baking soda?

4. Why doesn't baking powder need vinegar or sour cream to make batter rise, but baking soda does?

PART 1, CHAPTER 5: "CHEMICAL PARTNERS: WHO DOES WHAT TO WHOM"

Precipitation reactions happen in solutions, when two components meet that like each other better than they like water molecules; this new compound falls out of solution, or *precipitates*. Precipitation reactions are **recombination** reactions, that is, reactions in which compounds swap partners; but this is only because the two components of the *precipitate* needed other things to balance their electrical charge when they were dissolved.

For example, consider the reaction between copper sulfate and sodium bicarbonate (forming the "blue blob" in the demonstration).

- Copper sulfate ($CuSO_4$) dissolves in water because copper ions + water and sulfate ions + water are more stable than copper ions + sulfate ions in the solid. You need both positive ions (copper) and negative ions (sulfate) to have electrical neutrality. A solution of copper sulfate contains copper ions Cu^{2+} and sulfate ions SO_4^{2-}.
- Sodium bicarbonate (baking soda, $NaHCO_3$; "HCO_3" is the bicarbonate ion, charge -1) dissolves in water because sodium ions + water and bicarbonate ions + water are more stable than sodium ions + bicarbonate ions in the solid. You need both positive ions (sodium) and negative ions (bicarbonate) to have electrical neutrality.

- o Carbonate ions are always present in a solution of bicarbonate ions because bicarbonate can lose H⁺ to water:

$$HCO_3^- + H_2O \rightarrow CO_3^{2-} + H_3O^+$$

  or to itself:

$$2\,HCO_3^- \rightarrow CO_3^{2-} + H_2CO_3$$

  Thus, a solution of sodium bicarbonate contains all of the following: sodium ions Na^+, bicarbonate ions HCO_3^-, and tiny amounts of carbonate ions CO_3^{2-}, hydronium ions H_3O^+, and carbonic acid H_2CO_3. These ions and molecules are surrounded and stabilized by water molecules.

- When a solution of copper sulfate and a solution of sodium bicarbonate are mixed, copper carbonate precipitates because the marriage of copper ions + carbonate ions in the solid state is more stable than copper ions + water and bicarbonate ions + water. What remains in water solution is sodium sulfate and carbonic acid (or H_3O^+ and HCO_3^-), simply because those are the things that are left over from the original positive-negative ion pairs.

- The reaction is conventionally written

$$CuSO_4(aq) + 2\,NaHCO_3(aq) \rightarrow CuCO_3(s) + Na_2SO_4(aq) + 2\,H_2CO_3\,(aq)$$

  where *(aq)* indicates that the compound is in *aqueous solution* (dissolved in water) and *(s)* indicates that the compound is a solid.

- A more realistic representation is the following, since dissolved ions are separated from their opposite numbers:

$$Cu^{2+}(aq) + HCO_3^-(aq) \rightarrow CuCO_3(s) + H^+(aq)$$

  The other ions are *spectators* that do not participate. We have not stated "H_2O" explicitly because "*aq*" implies water solution. (We will see more of "balanced equations" in Chapter 7.)

- **ALL** the compounds described in this example are **salts**, that is, compounds composed of charge-balanced combinations of ions, with the exception of carbonic acid, H_2CO_3.
- In the demonstration, notice that the reaction fizzes. This is because carbonic acid decomposes into carbon dioxide gas and water:

$$H_2CO_3 \rightarrow H_2O + CO_2.$$

Electronegativity is the tendency of elements to attract electrons in bonds. Elements with high electronegativity tend to form negative ions, while those with low electronegativity tend to form positive ions. We will see more of this in the discussion of Chapter 6.

Molecular polarity

As we will see in Chapter 6, ionic bonds (an attraction between two ions) are one end of a continuum of *polarity* (charge separation) in chemical bonds. (At the other end are bonds with no polarity at all.) As we will see in Chapter 6, not only the presence of polar bonds but also molecular shape affects whether a molecule is polar.

Water is a polar molecule because its polar bonds (between oxygen and hydrogen) are arranged in a V-shape, with hydrogen atoms at the top of the V and oxygen at the bottom. The higher electronegativity of oxygen makes electrons concentrate at the oxygen end of the water molecule. This makes the oxygen end negatively charged, and the hydrogen end positively charged.

When water **solvates** ions (the noun is **solvation**), it surrounds them in such a way that the negative end of the surrounding water molecules points toward positive ions, and the positive end of the surrounding water molecules points toward negative ions. Often, this **solvation** arrangement is more stable than the interactions between ions in the solid – but not always, and that is the reason for *precipitation* reactions.

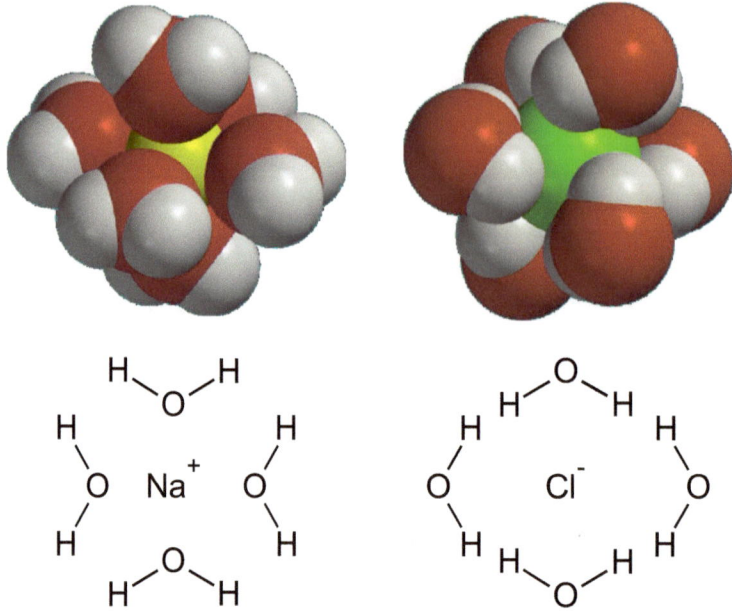

Sodium cation, Na⁺
solvated by water molecules

Chloride anion, Cl⁻
solvated by water molecules

Images © Daniel Berger

Be careful to understand the distinctions made in the text:
- Solubility varies with the properties of the solvent: oil vs. water, vinegar vs. water.
- Several examples of precipitation reactions are given.

Hard and soft water

Don't forget to study the distinction between hard water and soft water, which is discussed in the textbook.

More about hard and soft water:

- **Soap** forms a scum with hard water, but **detergent** doesn't.
- **Soap** contains sodium salts of fatty acids, that is, sodium carboxylates. Carboxylate ions are an organic variant on carbonate ion – limestone, which is only very slightly soluble in water, is calcium carbonate – and these fatty acid

carboxylates form a precipitate ("soap scum") with calcium ions, leaving their sodium ions behind in solution.
- **Detergent** contains other things that prevent the formation of soap scum.
    - Older detergents used conventional soap in combination with phosphate salts. Phosphate ions form water-soluble complexes with calcium ions, preventing them from forming soap scum. The detergent industry calls chemicals that trap calcium ions *builders*. Phosphates are builders.
    - Newer detergents avoid phosphates by using compounds similar to but different from soap. Soap substitutes include sulfate-based fatty acid salts that don't precipitate with calcium (such as "sodium lauryl sulfate," commonly seen in personal care products; sodium lauryl sulfate is a sulfate derivative of the fatty acid lauric acid), or so-called "non-ionic surfactants" which don't have any ions in their structures and so can't form precipitates with calcium.
    - Still other non-phosphate detergents use conventional soap with non-phosphate builders to trap calcium ions.
    - We will discuss surfactants in more detail in Chapter 6.

Phosphates in detergent are a big problem because phosphates are also excellent nutrients for plants, and when flushed into waterways will cause algae blooms, which choke other marine life.

Vocabulary

You should write your own definitions for these words, based on the textbook and this outline:

Precipitation

Aqueous solution

Salt

Solubility and insolubility (the adjectives are "soluble" and "insoluble")

Electronegativity

Hard water

Soft water

Builder

Questions

1. Define "electronegavity" and tell what it means for a chemical bond between two atoms.

2. If a molecule is "polar," what does that mean? What has to be true for a molecule to be polar?

3. Define "solvation" and describe how water solvates (for example) sodium cations or hydroxide anions.

4. Something that is "insoluble" will still dissolve. Explain this.

5. Often, when two soluble compounds are mixed a precipitate is formed. Describe, on the atomic/molecular level, what happens in such a reaction.

6. Describe the difference between hard and soft water (a) in terms of the ions dissolved in them and (b) in terms of how they interact with soap.

7. What is "soap scum"?

8. Describe how detergent differs from soap. Include information on how each interacts with ions dissolved in water.

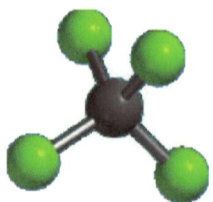

PART 1, CHAPTER 6:
"THE TIE THAT BINDS, THE CHEMICALS THAT BOND"

A note: This is longer than usual because bonding theory is my specialty. What I expect you to know is summed up in the vocabulary list. D.B.

Types of bonds

Recall that electronegativity is the tendency of an atom to hold electrons tightly. Chemical bonds consist of electrons distributing themselves between atomic nuclei so that everything is attracted to everything else. Therefore we expect bond types to differ as the atoms involved hold their electrons more tightly or loosely.

We've seen what an *ionic bond* is: it's the situation in which two or more ions are held together by electrostatic attraction, positive charges to negative charges. This can happen in a number of different ways; you don't need to have just binary ionic compounds like sodium chloride or calcium carbonate. Natural minerals can range all the way from simple combinations like silicon dioxide, SiO_2 (quartz), or calcium carbonate, $CaCO_3$ (limestone or calcite), to dolomite, $CaMg(CO_3)_2$ and beryl, $Al_2Be_3Si_6O_{18}$, to minerals with naturally varying composition such as olivine, $(Mg,Fe)_2SiO_4$, or titanium dioxide, which averages about $TiO_{1.8}$. *Of course, not all these minerals are "ionic compounds."*

In an ionic bond, electronegativities of the elements involved are different enough that electrons are shared very unequally, so much so that it's an excellent approximation to say that the atoms involved are permanently ionized.

On the other hand, in a bond between atoms of the same element, electrons are shared completely equally. There are two ways in which this can be done:

- **Covalent bonds.** When an element has a high electronegativity, only enough electrons are shared to form tight links between pairs of atoms. Each atom may have more than one pairwise electron-sharing interaction or "bond," each with a different neighbor; but the valence electrons involved in each bond are held tightly between each pair of atoms, like the rods in a Tinkertoy® model. This means that the electrons cannot flow, and such an element is an electrical insulator. Covalent bonds can also occur between atoms of different elements.

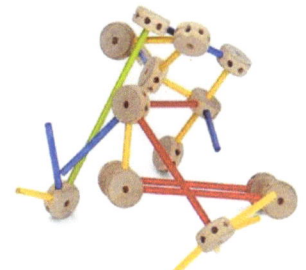

Image: Wikimedia

- **Metallic bonds.** When an element has low electronegativity, valence electrons are shared more loosely, so that they form a sort of continuous glue that holds the atoms together, like the glue in chip board. Some of the electrons, because they are not localized between pairs of atoms, can flow through the body of the material, making it electrically conductive. Metallic bonding can also occur between atoms of different elements.

Image: Wikimedia

It should be understood that these three types of bonding – ionic, covalent and metallic – shade into one another. Considering a bond between a pair of atoms:

- The higher the **difference** in the electronegativities of the two atoms, the more ionic the bond will be, that is, the more unequally electrons are shared between atoms.
- The *higher* the **average** electronegativity of the two atoms, the more tightly the electrons will be held between them and the more covalent the bond will be.

35

- The *lower* the <u>average</u> **electronegativity** of the two atoms, the less tightly electrons will be held between the two atoms and the more <u>metallic</u> the bond will be.

This can be represented by a triangular diagram, such as this one from Meta-Synthesis.com:

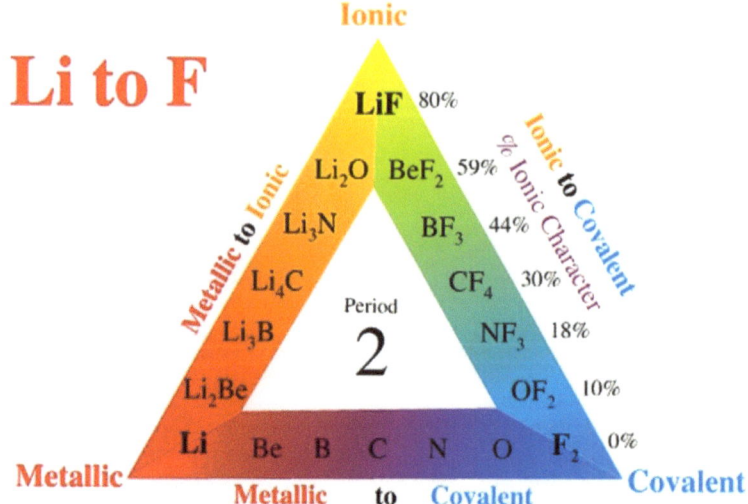

This diagram represents the types of bonding found in the second period, lithium to fluorine. In the diagram, the horizontal dimension is **average electronegativity** and the vertical dimension is **difference in electronegativity.**

Using the periodic table to predict bonding

Valence electrons and octets

As we saw in Chapters 2 and 3, the periodic table can be used to determine how many valence electrons an element has. This, with the octet rule, can be used to predict the charge of an atomic ion. **It can also be used to determine how many covalent bonds can be formed by a non-metal,** as discussed briefly in Chapter 2.

Remember that the number of valence electrons that atoms in any family have is given by either the family number (if less than or equal to 10) or the family number minus 10 (if greater than 10). Thus, lithium (Family 1) has one valence electron; beryllium (Family 2) and zinc (Family 12) have two; scandium (Family 3) and boron (Family 13) have three; titanium (Family 4) and carbon (Family 14) have four; and so on.

The number of covalent bonds an atom is normally able to form (its *valence*) is the same as the number of electrons it needs to gain or lose to get to the nearest octet, but in any case an element can seldom form more covalent bonds than it has valence electrons. Thus, hydrogen and fluorine can each form only one covalent bond, because they need to *gain* just one electron to get to the nearest octet, and they have enough valence electrons to do so; boron can form only three bonds because it has only three valence electrons, even though four bonds are needed for it to have an octet.*

Notice that you can have more than one bond between the same pair of atoms. A simple example is the most common form of elemental oxygen, O_2. Each oxygen atom wants to form two bonds, and rather than linking up in a chain, $-O-O-O-$, they pair up, $O=O$, to form a double bond. Carbon dioxide does the same thing: carbon wants four bonds, each oxygen wants two, so the carbon forms a double bond to each oxygen atom: $O=C=O$.

There are exceptions to the octet rule, called "hypervalence." Atoms below the second period (notably sulfur and phosphorus, but also several other non-metallic elements) can form as many covalent bonds as they have valence electrons. Simple examples include PF_5 and SF_6, in which phosphorus and sulfur form five and six covalent bonds respectively.

In ions, things are looser. While elements cannot have *oxidation states* higher than their number of valence electrons, they can have any number of covalent bonds, subject only to the oxidation state rule just given and the number of atoms they can fit around themselves. For example, the hexafluorophosphate ion PF_6^- has phosphorus with an oxidation state of +5, but forming six covalent bonds, one to each of the fluorine atoms. This works because the negative charge provides an extra valence electron for phosphorus.

The situation is similar in the perchlorate, perbromate and periodate anions (ClO_4^-, BrO_4^- and IO_4^-). In each of these, the

---

* Boron can form four bonds if it is negatively charged so that it has four valence electrons. Borate anions are more common than neutral compounds of boron, because in a borate ion, boron has an octet.

halogen atom has an oxidation state of +7, but forms four double bonds, one to each oxygen atom, for a total of eight.*

Class of bond vs. placement and distance on the periodic table

Bonds between two metals will obviously be metallic. Bonds between two non-metals, or two metalloids, or between a non-metal and a metalloid, will be covalent. Ionic bonds usually form between elements that have a large horizontal separation on the periodic table, for example, calcium and oxygen or sodium and chlorine which are far apart; normally ionic bonds are found in compounds containing both a metal and a non-metal.

But bond types can vary all over the bonding triangle (ionic, covalent, metallic). Normally we only consider ionic and covalent bonds as the endpoints of the bonding continuum. But remember that there's also a continuum between metallic and ionic, and some bonds can have characteristics of all three ideal types.

Conventionally, in the transition from pure covalent bonds (such as F_2 or H_2) or pure metallic bonds to ionic bonds (no bond is **purely** ionic, though cesium fluoride comes pretty close), about 80% of the variation falls under the heading of **polar covalent**. This is because of the way the bonds behave.

"Behavioral" definitions of bonding

Ionic compounds have ions stacked like oranges in a crate. Each ion is surrounded by ions of opposite charge, and all of the nearest-neighbor ions are the same distance away. The number of nearest neighbors is dependent on the sizes of the ion and its neighbors. Ionic compounds tend to be brittle, and to have moderately low melting points but extremely high boiling points. Interestingly enough, ionic compounds are insulators in the solid, but conduct electricity very well

---

* The more knowledgeable reader will recognize that this is a slight oversimplification. Each oxygen-halogen bond in perchlorate or perbromate ion is "1¾ bonds" because of averaging: of the four, one is single, but since all the halogen-oxygen bonds are equivalent they have an average bond order of 1.75.

when molten because ions in a molten salt are able to move freely. There will be more about this below.

Metallic materials, like ionic compounds, have atoms stacked like oranges in a crate, and the number of nearest neighbors depends largely on the size of the atom and its neighbors. However, metals are ductile and conduct electricity well in both the solid and liquid phase. Electrical conductivity and ductility are due to the similarity of electrons in a metallic bond to stiff glue in which the atoms are embedded: if the material is bent or stretched, the electrons are able to readjust; and some of the electrons are able to move more-or-less freely.

The majority of compounds between two different elements are held together by covalent bonds. Covalent bonds are characterized typically by the formation of well-defined molecular structures, with molecules separated from other molecules by distances larger than the separation of atoms within the molecules. There are a few exceptions to this rule: carbon (in the form of diamond), crystalline silicon and germanium, and certain compounds such as boron nitride, silicon dioxide and gallium arsenide have extended – formally infinite – network structures in which each atom is covalently bound to its nearest neighbors. All other non-metallic elements and compounds form discrete molecules of greater or lesser (but finite) size.

Covalent compounds or elements, like ionic ones, are poor electrical conductors in the solid; but covalent compounds are also poor conductors in the liquid phase because they do not form ions when they melt. Some covalent compounds, notably acids and ammonia, form ions when dissolved in water, but this is a *chemical* process in which a neutral compound is transformed into ions; see Chapter 4.

Polyatomic ions such as sulfate (SO_4^{2-}), carbonate (CO_3^{2-}) or ammonium (NH_4^+) are held together by covalent bonds.

Electrical conduction

Electricity is conducted by mobile, electrically charged particles of any type. We normally think of electrons as the carriers – and they do carry electricity through metals – but molten salts also conduct electricity because in the liquid phase the ions of a salt can move freely past each other.

When an ionic compound is dissolved in water, its ions become surrounded by water molecules and can move freely. Therefore, ionic compounds conduct electricity when dissolved in water. Such compounds are known as **electrolytes**. Certain covalent compounds, such as acids and ammonia, also generate ions when dissolved in water (see Chapter 4) and are therefore electrolytes. Electrolytes will be further discussed in Chapter 18.

You should carefully read the discussion of semiconductors, including **valence bands** and **conduction bands** (of electrons within a solid), and the difference and similarity between **electrons** and **holes**, toward the end of Chapter 6.

Vocabulary

You should write your own definitions for these words, based on the textbook and this outline:

Electronegativity

Polar

Non-polar

Ionic bond

Covalent bond

Metallic bond

Octet rule

Polar covalent bond

Electrical insulators, electrical conductors, electrical semiconductors

Electrolytes

Valence band

Conduction band

Doping

Electrons and holes

Questions

1. "All bonds between atoms involve shared electrons." True or false? Defend your answer.

2. Describe the differences and similarities between ionic, covalent and metallic bonds. What does it mean to talk of "the bonding continuum"?

3. Find at least two pairs of elements on the periodic table that are expected to form an ionic bond.

4. Find at least two pairs of elements on the periodic table that are expected to form a covalent bond.

5. Find at least two pairs of elements on the periodic table that are expected to form a metallic bond.

6. Why doesn't hydrogen form ionic bonds as a cation? That is, why isn't *e.g.* hydrogen chloride an ionic compound?

7. Why doesn't pure silicon conduct electricity? What is doping, and how does doping turn silicon into a semiconductor?

PART 1, CHAPTER 7: "STICKING TO PRINCIPLES"

All images in this chapter © Daniel Berger

Items to pick up on in the reading:

The law of definite proportions

Elements combine to form compounds in specific mass ratios. These mass ratios were how atomic mass was first determined (now it's done with mass spectrometry).

This law has consequences: you can have *different compounds* made up of the *same elements* in different proportions. The example given in your text is water (hydrogen oxide, H_2O) vs. hydrogen peroxide (H_2O_2). Another simple example is the different types of copper ore: chalcocite is Cu_2S and covellite is CuS. Iron has two different oxides, FeO (wüstite) and Fe_2O_3 (hematite).

My subfield, organic chemistry, shows – as a conservative estimate! – several million examples of different compounds made of the same elements (see Section 2, Chapter 1). There are many times more known compounds containing carbon than there are compounds *not* containing carbon, using all the rest of the periodic table in any combination whatsoever.

The entire family of "hydrocarbon" compounds is made of only hydrogen and carbon, in myriads of different arrangements and proportions. What comes out of an oil well is almost exclusively made of the elements hydrogen and carbon, but contains hundreds or thousands of different compounds made of those two elements, with formulas ranging from CH_4 to $C_{30}H_{62}$ and beyond.

Condensed formulas

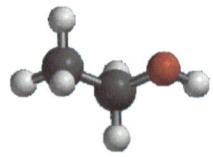

A *condensed formula* is simply a list of the elements in a compound, and the number of "atom units" of each element (for example, in water there are two atoms of hydrogen for every atom of oxygen, and so the formula is H_2O). Sometime the condensed formula can give structural information; the formula for ethyl alcohol can be written either C_2H_6O or CH_3CH_2OH, with the second condensed formula giving structural information similar to that in the ball-and-stick model shown here.

Conventionally, we write condensed formulas using a number of sometimes mutually contradictory rules (Notice that I use acetic acid under all three rule sets, below).

1. For ionic compounds or other compounds of metals, metal atoms or cations are listed first: for example, LiBr, not BrLi. Acids are listed with the acidic hydrogen atom(s) given first: hydrochloric acid is HCl, sulfuric acid is H_2SO_4 and **acetic acid** is $HC_2H_3O_2$: three of the hydrogen atoms in acetic acid are **not** acidic.

2. In simple structures, or in condensed formulas that give structural information, the central atom of a group is given first, followed by the surrounding atoms. For example, ammonia is NH_3, ethyl alcohol is CH_3CH_2OH (three central atoms in a row, with surrounding hydrogens) and acetic acid is CH_3COOH. **Important exception:** compounds of hydrogen with Group 16. Water is H_2O, not OH_2, and hydrogen selenide is H_2Se, not SeH_2. For these compounds, we are following Rule 1.

3. In organic chemistry (the chemistry of carbon compounds), condensed formulas that do not try to convey structural information give carbon first, then hydrogen, then any other elements in alphabetical order. For example, using this rule ethyl alcohol is written C_2H_6O, acetic acid is written $C_2H_4O_2$ and acetaminophen is written $C_8H_9NO_2$. This is sometimes called

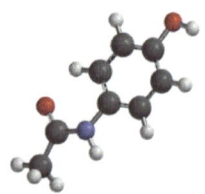

acetaminophen

the *Chemical Abstracts* convention, because it is the system used by the *Chemical Abstracts* database.

Isomers

Compounds that have the same formula (meaning that they have the same number of the same kinds of atoms) but different structures are called *isomers*. This phenomenon is particularly common in organic chemistry, though a number of inorganic compounds also exhibit *isomerism*. Your text lists the three isomers of CHNO:

- Fulminic acid, H−C≡N−O
- Cyanic acid, H−O−C≡N
- Isocyanic acid, H−N=C=O

As another example, both ethyl alcohol and dimethyl ether have the formula C_2H_6O, but they have different structures:

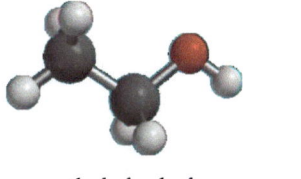 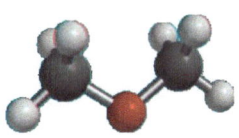

ethyl alcohol **dimethyl ether**

For isomers, Rule 2 is important. To distinguish the isomers, we write ethyl alcohol as CH_3CH_2OH or C_2H_5OH, and dimethyl ether as CH_3OCH_3 or $(CH_3)_2O$.

Another common example from organic chemistry: commercial butane (used in lighters and backpacking stoves) is a mixture of C_4H_{10} isomers:

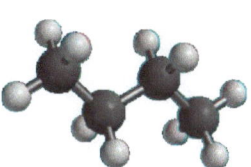 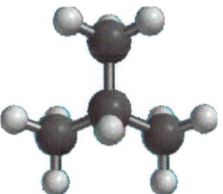

45

Conservation of mass

What we saw in the "Gemstone Chemistry" demonstration was the conservation of mass: mass is neither created nor destroyed in chemical reactions. We expect that, if we can account for everything, the material we put in will weigh the same as the material we get out of a chemical reaction. For example, in the neutralization of vinegar by baking soda, the products should weigh the same as the reactants:

$$HC_2H_3O_2 + NaHCO_3 \rightarrow NaC_2H_3O_2 + H_2O + CO_2\uparrow$$
(the up-arrow indicates a gaseous product)

The example used at the end of Chapter 7 is chemical engineering, and the specific chemistry discussed is that of the *chlor-alkali process*.

$$2\ H_2O + 2\ NaCl \rightarrow 2\ NaOH + Cl_2\uparrow + H_2\uparrow$$

In the chlor-alkali process, strong brine (salt water) is electrolyzed, producing three useful products: lye (NaOH), chlorine and hydrogen. These products are always produced in the same ratio, because of the atom-balance or *stoichiometry* of the process. (More about stoichiometry in Chapter 9.)

This can be a financial problem for chlor-alkali plants, because the price of lye and the price of chlorine don't usually track one another well. In order to make lye that the market is demanding, the producer might need to dump chlorine below cost, or vice versa. (The market for hydrogen is robust, so no worries there.)

Vocabulary

You should write your own definitions for these words, based on the textbook and this outline:

Law of definite proportions

Condensed formula

Isomers

Conservation of mass

Questions

1. Which of the following condensed formulas gives the most structural information? What structural information *is* given? All formulas are for the same compound, butyric acid (the compound responsible for "stinky armpit odor").

    $C_4H_8O_2$ $HC_4H_7O_2$ $CH_3CH_2CH_2COOH$

2. What is *always* true of *isomers*?

3. For each of the chemical reactions shown below, balance the reaction expression.

    $HC_2H_3O_2 + NaHCO_3 \rightarrow NaC_2H_3O_2 + CO_2 + H_2O$ (mixing vinegar and baking soda)

    $HClO + HCl \rightarrow H_2O + Cl_2$ (you shouldn't mix bleach with toilet cleaner)

    $C_4H_{10} + O_2 \rightarrow CO_2 + H_2O$ (burning butane)

    $Zn + MnO_2 \rightarrow ZnO + Mn_2O_3$ (discharging a flashlight battery)

    $PbSO_4 + H_2O \rightarrow Pb + PbO_2 + H_2SO_4$ (charging a car battery)

    $C_6H_{12}O_6 + O_2 \rightarrow CO_2 + H_2O$ (metabolizing sugar)

    $CO_2 + H_2O + Na_2CO_3 \rightarrow NaHCO_3$ (carbon dioxide dissolving in the ocean)

PART 1, CHAPTER 8: "SLIPPING AND SLIDING, INTERMOLECULARLY"

> You may have noticed that the authors of *The Joy of Chemistry* have misread the quote from *The Picture of Dorian Gray*. The affinity Oscar Wilde was describing was, of course, not between the "chemical atoms" of the painting and other chemical atoms of the painting, but between the atoms of the painting and Dorian Gray's soul. I recommend the tale; it's one of my favorites.

There's not a great deal for me to add to Chapter 8. A few things to emphasize:
- How surfactants work. This is begun in the discussion of Demonstration 8, and continued at the end of the chapter.
- The difference between <u>inter</u>molecular forces and <u>intra</u>molecular forces.
- Types of intermolecular forces:
    o Ion-ion interactions ("ionic bonds") are not normally considered "intermolecular."
    o Ion-dipole interactions are between ions and polar molecules. When ions are dissolved in a solvent, it is this sort of interaction that keeps them in solution.
    o Dipole-dipole interactions, between two polar molecules
    o Dispersion forces (London forces). These involve *induced dipoles*, also called *temporary dipoles*. How this happens is well-explained in your text.
- Note the contribution of electronegativity to bond polarity, and the importance of molecular shape to whether a molecule with polar bonds is actually polar (e.g. H_2O vs. CO_2).

- What is a "hydrogen bond"? This is well-explained in your text.
- As always, pay attention to the concepts introduced in the "For Example" section; this is how you find out how what you've learned applies to everyday life.

More about London (dispersion) forces

Many students have difficulty with the concept of London forces. Their source is explained well, if perhaps too briefly, by Cobb and Fetterolf. However, there is an analogy which may be helpful.

London forces work by surface contacts between molecules. The more surface area a molecule has, the more opportunity it has to use part of that surface for a weak, dispersion-force interaction with some other nearby molecule. If there are a lot of molecules nearby (as in a liquid or solid), these very small forces can add up rapidly – though never to the sort of attractions you get between, say, two water molecules.

London force interactions in a liquid can be likened to a colander full of pasta that's been well rinsed. Long spaghetti sticks to itself fairly well, even after all the starch has been washed away: each strand has lots of surface area in contact with lots of other spaghetti strands, which in turn are in contact with yet more, and so on.

On the other hand, elbow macaroni, with its much lower surface area, has fewer opportunities to contact neighboring pieces of macaroni. This means that it doesn't stick to itself well if it's been well rinsed.

Vocabulary

You should write your own definitions for these words, based on the textbook and this outline:

Intermolecular

Intramolecular

Dipole

Dispersion forces

London forces

Hydrogen bonds

Lubrication

Adhesion

Capillary action

Surfactant

Explain how siphoning works

Questions

1. What is a surfactant? How does it facilitate interactions between two molecules that would otherwise not want to interact?

2. Describe how water molecules *solvate* (that is, dissolve) an ionic compound. (See also Chapter 5)

3. What is bond polarity, and how does it relate to molecular polarity?

4. Describe the similarities and differences between a dipole-dipole interaction and a hydrogen bond.

5. Describe how London forces are generated. Why are London forces so weak?

PART 1, CHAPTER 9: "CONCENTRATION—ON BEING ALONE TOGETHER"

We normally think of "solutions" as being "solutions **in water**" (*aqueous* solutions). But there are a number of commonly-encountered solutions that are not liquids, and are not aqueous. Air is a solution of oxygen and other gases in nitrogen, and metal alloys such as brass, bronze and steel are also solutions.

Furthermore, it's not only solids that dissolve in liquids: gases will also dissolve. The pressure of a gas above a liquid effects the concentration of that gas in that liquid (higher pressure leads to higher concentration, lower pressure to lower concentration). This is why carbonated beverages are pressurized with carbon dioxide when they are bottled (pressurization raises the concentration of carbon dioxide in the beverage), and why they fizz when opened (opening allows the extra carbon dioxide to escape, by lowering the *partial pressure* of carbon dioxide above the beverage).

The mole as a chemical unit

The *mole* is the molecular mass or formula mass, stated in grams. It so happens that there are 6.0221415×10^{23} or 602,214,150,000,000,000,000,000 molecules per mole, but that number is not important! The important concept is *the formula mass in grams*.

For example, the molecular mass of water (H_2O) is equal to twice the atomic weight of hydrogen (about 2 atomic mass units for two hydrogen atoms) plus the atomic weight of oxygen (about 16 atomic mass units for one oxygen atom), because water molecules consist of two atoms of hydrogen and one of oxygen. The molecular mass of water therefore comes to 2 × 1 + 16 = 18 atomic mass units. Thus, one mole of water has a mass of 18 grams, and this is referred to as the *molar mass*. See how it works?

To determine how many moles of water are in 5 grams of water, divide the mass of water you have (5 grams) by the molar mass (18 grams per mole). 5 ÷ 18 = 0.278; you therefore have 0.278 moles of water in 5 grams of water.

We will see later how the mole allows us to do practical chemistry.

Concentration: how much you have dissolved in how much solution

Concentration can be expressed in a number of ways ("proof," for example, is twice the percentage of ethyl alcohol in water), but the most chemically useful way to state concentration is in terms of moles. Because it's easy to measure liquids by volume, the concentrations of things dissolved in liquids are usually expressed as "moles per liter."

A concentration of xx moles per liter means that if you measure out 1 liter of the solution, you have xx moles of solute. Concentration as moles per liter is calculated as (moles of solute) ÷ (liters of solution). Thus, if you dissolve 2 moles of sulfuric acid in enough water to make a total volume of one liter, the concentration is 2 moles per liter.

This is NOT the same as dissolving two moles of sulfuric acid in one liter of water! Almost always, the total volume of a solution is significantly different than the volume of the solvent alone (unless the solution is **very** dilute).

You can try this at home: put ¼ to ½ cup of sugar in a 2-cup measuring cup, and – while keeping track of the amount of water you use – add enough water to make two cups of solution. You will need to stir a good bit (and perhaps heat, gently) to dissolve the sugar. Did you use two cups of water, or some smaller amount? What total volume do you get if you dissolve ¼ cup of sugar with an entire cup of water?

Other concentrations:

Mixtures of gases are normally stated in terms of percentage by volume, since every gas molecule, whatever it may be, occupies the same volume of space (on average) as every other gas molecule. We will explore this further in Chapter 10.

Acid and base strength is stated in terms of pH which (crudely) is minus-one times the common logarithm* of the concentration of hydronium ion, in moles per liter. For example, consider a solution of acid in water that contains 0.001 moles of hydronium ion per liter. The common logarithm of 0.001 (or 10^{-3}) is -3, and so the pH is equal to $+3$. A more concentrated solution of acid, with 5 moles of hydronium ion per liter, has a pH of -0.7, since the common logarithm of 5 (or $10^{0.7}$) is $+0.7$.

Bases have pH higher than 7, since the concentration of hydronium ion in pure water is 10^{-7} moles per liter. Water *autodissociates* to that extent, separating into hydronium ion and hydroxide ion in equal amounts. Extra hydroxide ion soaks up most of that small amount of hydronium ion, lowering its concentration even further.

---

* The "common logarithm" of a number is the exponent to which 10 must be raised to get the number. For example, since $100 = 10^2$, the common logarithm of 100 is 2. We symbolize that as log 100 = 2.

Vocabulary

You should write your own definitions for these words, based on the textbook and this outline:

Solution

Solute

Solvent

Aqueous solution

Molecular mass

Mole (as a unit in chemistry)

Concentration

Common logarithm (usually just called "logarithm")

pH, in terms of the concentration of hydronium ion

Exercises

Formula weight calculations

Calculate the molar masses, also called formula weights, of each of the following compound. The formula is given to you. Don't forget to use the correct unit of mass!

a. Carbon disulfide, CS_2

b. Ethanol, C_2H_5OH

c. Boric acid, H_3BO_3

d. Octane, C_8H_{18}

e. Calcium carbonate, $CaCO_3$

f. Ammonium phosphate, $(NH_4)_3PO_4$

Mole calculations

Calculate the number of moles in the given amount of each of the compounds below.

20 grams of carbon disulfide, CS_2

454 grams of ethanol, C_2H_5OH

28 grams of boric acid, H_3BO_3

3200 grams of octane, C_8H_{18}

250 grams of calcium carbonate, $CaCO_3$

500 grams of ammonium phosphate, $(NH_4)_3PO_4$

PART 1, CHAPTER 10:
"IT'S A GAS"

Cobb and Fetterolf do a good job of explaining most of the concepts in this chapter.

Notice that, to describe a gas completely, we need to know four things:
    a) Its temperature
    b) Its pressure
    c) The volume it occupies
    d) The amount of gas present

Or, laid out individually,

Temperature: a gas' pressure increases with its temperature if the volume is held constant.	$P \propto T$
Temperature: the volume a gas occupies increases with its temperature if pressure is held constant.	$V \propto T$
Pressure: a gas' temperature increases as pressure rises, and vice versa.	$T \propto P$
Pressure: a gas' volume decreases as external pressure rises, and its pressure decreases as its volume increases.	$\frac{1}{V} \propto P$
Volume: a gas' temperature decreases as it expands, and rises as it is compressed.	$\frac{1}{T} \propto V$
Volume: a gas' pressure decreases as it expands, and rises as it is compressed.	$\frac{1}{P} \propto V$
Quantity: as the number of particles rises, the pressure of a gas rises if volume is held constant.	$P \propto n$

Quantity: as the number of particles rises, the volume of a $V \propto n$ gas rises if pressure is held constant.

This set of relationships corresponds to the Ideal Gas Law, $PV \propto nT$, or $\frac{PV}{nT} = R$, where R is "the gas constant."

These things can be accounted for by *kinetic molecular theory*, the model of gases (and other materials) as being composed of small particles in constant motion. This is explained well in the text.

There is **one inadequate explanation in the text**, however. We are not crushed when we exhale *because our internal pressure remains always equal to that of the outside air*. When we exhale, only the air equal to the reduced volume of our chests (from breathing motion) is expelled. Enough air remains in our lungs to keep them from collapsing under the external pressure.

In the "For Example" section, flatulence is discussed and the way odors spread is noted. This will be taken up again in Demonstration 11, "It's In the Air."

The Kelvin temperature scale, and how it relates to the Celsius scale

Real gases at "normal" temperatures and pressures (that is, similar to those we have on Earth) behave almost exactly like an ideal gas, that is, one that has zero molecular size and zero intermolecular interactions. This means that we can use the temperature-volume relationships for real gases to estimate the value of absolute zero.

From a standard table of densities,[*] we obtain values for the density of air at different temperatures and pressures. We can convert densities to volumes by simply taking their reciprocals (volume per unit mass), and convert Fahrenheit temperatures to Celsius temperatures using $C = 5/9 (F - 32)$. When we plot the volume of a gas vs. its Celsius temperature at a constant pressure, and extrapolate to zero volume, we get a value of $-273°C$.

---

[*] "Air – Temperature, Pressure and Density," The Engineering ToolBox. www.engineeringtoolbox.com/air-temperature-pressure-density-d_771.html (accessed 10 October 2013).

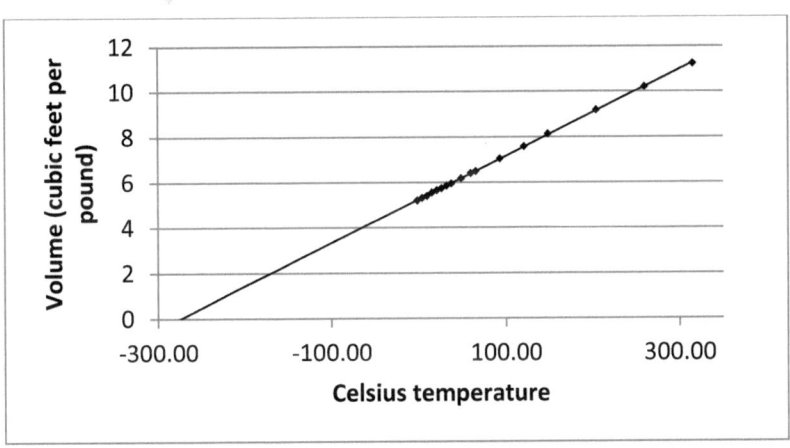

Air volume vs. temperature at a pressure of 30 pounds per square inch (2 atmospheres)

Notice, from the graph, that volume has a direct, linear relationship with temperature; in other words, the volume is directly proportional to the value of the temperature.

The Kelvin temperature scale – also known as the "absolute temperature" – is defined as having units equal to 1°C, with its zero value at the temperature at which an ideal gas's volume is equal to zero (−273°C). Thus, to convert between Celsius and Kelvin, we just add 273 to the Celsius temperature value.

How to solve proportionality questions using the properties of gases

Recall that

- The volume of a gas is directly proportional to its temperature and inversely proportional to its pressure: $V \propto \frac{T}{P}$
- The temperature of a gas is directly proportional to its volume and its pressure: $T \propto VP$
- The pressure of a gas is directly proportional to its temperature and inversely proportional to its volume: $P \propto \frac{T}{V}$

This means that, for any two samples of a gas A and B,

$$\frac{P_A V_A}{T_A} = \frac{P_B V_B}{T_B}$$

so that if you know five of the values, you can solve for the sixth. Most of the Sapling Learning problems give you only "two values," but that means that the third is held constant; for example, **if you assume constant temperature**, you can determine the pressure under 10 feet of water by comparing the volume of a balloon at that depth to its volume in the air at 1 atmosphere of pressure. Temperatures are the same on both sides, and so they cancel:

$$\frac{P_A V_A}{T} = \frac{P_B V_B}{T} \text{ becomes}$$

$$P_A V_A = P_B V_B$$

NOTE THAT it is essential to use Kelvin (absolute) temperatures when working such problems! Celsius or Fahrenheit temperatures will <u>not</u> work!

Vocabulary

You should write your own definitions for these words, based on the textbook and this outline:

Diffusion

Temperature, also in terms of the kinetic molecular theory (KMT)

Pressure, also in terms of the KMT

Volume

Why we use moles in the Ideal Gas Law rather than grams (KMT will help with this)

Understand how the four (T, P, V and n) relate to each other in the Ideal Gas Law

Absolute zero

Kelvin temperature scale, also called the absolute temperature scale

Questions

1. Describe how the volume of a gas-filled balloon changes with changes in
    a. temperature.
    b. external pressure.
    c. the amount of gas in the balloon.

   Describe how each set of changes is explained by the kinetic molecular theory of gases.

2. Describe absolute zero, in terms of
    a. how the volume of a gas changes with temperature.
    b. the kinetic molecular theory of gases.

3. Why does a real gas condense to a liquid, or freeze to a solid, before reaching absolute zero?

PART 1, CHAPTER 11:
"WHEN GASES PUT ON AIRS"

As usual, we are largely noting important or missing points from Chapter 11.

Note the examples of *diffusion* in Experiment 11, "It's In the Air." The demonstration page on Moodle also shows how gas flows: it's a fluid.

Free radicals

Free radicals are the main drivers of gas phase chemistry. Examples:

- Combustion of hydrogen and natural gas, and also gasoline and other liquid hydrocarbons; in practice, the liquid doesn't burn; it evaporates and the vapor burns. Molecular bonds are broken by heat, and the resulting free radicals react violently with the rest of the fuel.
- Ozone generation by nitrogen oxides at ground level. Nitrogen oxides are themselves free radicals, and combine with oxygen to produce ozone and nitrogen molecules (N_2).
- Ozone generation and destruction in the upper atmosphere. Oxygen molecules (O_2) react with free oxygen atoms formed by solar ultraviolet light to form ozone (O_3). Other free radicals, also formed by solar ultraviolet, react with ozone to form O_2 again.

Because combustion is a free-radical process, it requires initiation by a spark. This is enough to generate a few radicals, and they get things going. This is our first example of an *activation energy*. More about that in Chapter 17.

Note the explanation of *quenching* and the explanation of why oxygen (O_2) is so reactive (it's a *diradical*).

Halogens such as fluorine and chlorine are even more reactive than oxygen, but it's because the F—F or Cl—Cl bond is so weak. It's easy to break, and when that happens, you are left with two F· or Cl· free radicals.

The **explanation** of why gunpowder is explosive (a solid substance giving gaseous products) is correct but **incomplete**. The reason gunpowder explodes, but powdered charcoal does not, is that gunpowder contains its own *oxidizer*: saltpeter (sodium nitrate) that liberates oxygen when heated. The oxygen surrounds and reacts with the powdered charcoal and sulfur in gunpowder, burning them very rapidly to carbon dioxide and sulfur di- and trioxide gases.

You can also cause explosions by mixing a fuel intimately with air:

- A bag of powdered charcoal will not burn well *unless you throw it up in the air*. That way, it's surrounded by oxygen. This is a situation analogous to gunpowder, in which the charcoal is mixed with a solid oxidizer.
- Dust explosions (in empty or part-empty grain bins that have a lot of grain dust in the air) occur because fine, combustible dust is intimately mixed with oxygen. The explosion of the *USS Maine*, that started the Spanish-American War, is thought by some to have been due to a coal dust explosion. Combustible dust mixed with air is an explosive situation, similar to...
- Internal-combustion engines spray a fine mist of fuel <u>mixed with air</u> into the cylinder. This burns explosively.
- Fuel-air explosives spray a fine mist of fuel into the air, then ignite it.

Gases are **dangerous** because

- Irritant or toxic gases such as carbon monoxide or chlorine flow like any other fluids and mix with the surrounding air by diffusion. Eventually this mixing will dilute them until they are no longer harmful (like pouring a cup of cyanide in the ocean), but that takes some time. Too much emission in one locality will cause serious problems if it isn't allowed the time to disperse; this can be seen in any large city, and even in Lima, Ohio.

- Excessive quantities of less-toxic or non-toxic gases that are not oxygen, such as carbon dioxide or nitrogen, can displace oxygen from the air, suffocating living things. This is especially dangerous with carbon dioxide, because it is heavier than air and collects in low places.* People inhaling the intoxicant nitrous oxide, or even the completely harmless gas helium, have died from suffocation because they have taken in so much not-oxygen that the oxygen in their lungs is displaced.

Catalysts in chemistry

Industrial gas-phase reactions normally take place under high pressure, often with high heat, and typically require a *catalyst* – something that facilitates the reaction. Several examples are discussed in your text:

- The Haber process,† which converts N_2 and H_2 into NH_3 using an iron oxide catalyst.
- The catalytic converter in your car, which converts unburned fuel C_xH_y, carbon monoxide CO, and nitrogen oxides NO_x into H_2O, CO_2 and N_2. This device uses platinum, palladium or rhodium catalysts, collectively known as precious-metal catalysts.
- The Fischer-Tropsch process, so-called in your text, begins with a combination of coal gasification and the water-shift reaction. Together these produce a mixture of carbon monoxide and hydrogen called *synthesis gas* or *syngas*. The Fischer-Tropsch reaction converts syngas into medium- to long-chain hydrocarbons and water, using a variety of metals as catalysts, notably cobalt and iron.

---

* Carbon dioxide is also toxic in large quantities because it lowers the pH of the blood (hyperacidosis) and because lower blood pH triggers a reflex in which people breathe faster, using up the available oxygen.
† For more, see Chemguide's article on the Haber Process, www.chemguide.co.uk/physical/equilibria/haber.html (accessed 10 October 2013).

- **It is essential to note** that a catalyst *will never cause a reaction that is not favorable* under the reaction conditions. What a catalyst *does* is lower the energy barrier (*activation energy*, Chapter 17) that must be surmounted for the reaction to happen.
    - In the Haber process, the reaction $N_2 + 3\,H_2 \rightarrow 2\,NH_3$ is exothermic (see Chapter 13) and therefore favorable.
    - In a catalytic converter, "burning" hydrocarbons, carbon monoxide and nitrogen oxides to carbon dioxide, water and nitrogen (N_2) is quite favorable. (Again, see Chapter 13.) This is similar to the reaction in gunpowder, in which nitrates, composed of nitrogen and oxygen, supply oxygen for the combustion of the rest of the gunpowder by breaking down to a mixture of N_2 and O_2.
    - In the Fischer-Tropsch process, the conversion of CO and H_2 into hydrocarbons and water is, once again, favorable and would happen eventually in nature under similar conditions. For example, the atmosphere of Jupiter has no free oxygen, carbon monoxide or carbon dioxide, because there is *so very much* hydrogen there that any O_2, CO or CO_2 that might be present is converted into CH_4 and H_2O.

We will discuss **activation energies** in more detail in Chapter 17.

For example: photosynthesis and respiration

Notice that, for electrons to go from point A to point B, there must be both a *source* and a *sink*:

- A *source* of electrons is something that is able to give up electrons.
- An electron *sink* is something that is able to absorb electrons.
- Redox reactions, of course, all have an electron source (the reducing agent) and an electron sink (the oxidizing agent).

You should recognize that photosynthesis is the reverse of *respiration*. All living things respire, including plants – which means that plants use oxygen just like animals. But when plants are exposed to light, they generate more oxygen than they consume: this is photosynthesis.

Vocabulary

You should write your own definitions for these words, based on the textbook and this outline:

Diffusion

Initiation of a reaction

Radical or "free radical"

Quenching (can be used for other types of reactions besides those with radicals, but that is rare)

Oxidizer

Catalyst

Respiration

Photosynthesis

Fermentation

Be able to explain the processes of...

A dust explosion

Why gunpowder is explosive (both the chemistry and the gas-physics)

The bare bones of the Haber process, and why it is so important

Where electrons go in photosynthesis

Why and how plants use oxygen

Questions

1. Explain why free-radical reactions tend to occur in the gas phase, rather than the liquid or solid phase.

2. Explain why gases are considered *fluids* like water or gasoline.

3. Explain why gasoline vapor burns more readily than liquid gasoline.

4. Define a *catalyst* and give two examples of processes that use catalysts.

5. At the beginning of the "For example" section on photosynthesis, it says that for electrons to flow, they need both a source and a place to go. How does this apply to redox reactions?

6. Is photosynthesis a redox reaction? Discuss how you can tell from the balanced reaction expression.
$$6\ CO_2 + 6\ H_2O \rightarrow C_6H_{12}O_6 + 6\ O_2$$

7. Is fermentation a redox reaction? Discuss how you can tell from the balanced reaction expression:
$$C_6H_{12}O_6 \rightarrow 2\ CO_2 + 2\ C_2H_5OH$$

Part 1, Chapter 12: "Crystal clear chemistry"

Images © Daniel Berger unless otherwise noted

There are a number of important concepts concerning chemical structures in Chapter 12. You will be expected to predict simple molecular geometries; see the pages that set out examples, linked from the Moodle site.

Molecules in 3-D

Molecules have 3-dimensional shapes. The shape can be the only difference between isomers:

- *Constitutional isomers* have the same atoms in a different order, like ethyl alcohol (CH_3CH_2OH) and dimethyl ether (CH_3OCH_3).
- *Stereoisomers* have the same atoms in the same order, but oriented differently in 3-D so that they have different shapes. There are two kinds of stereoisomer:
  - *Diastereomers*, stereoisomers that are **NOT** mirror images, like galactose and glucose.

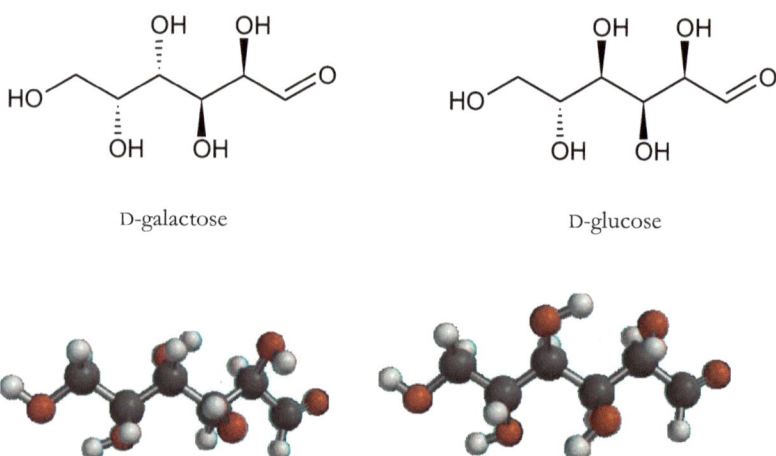

- Enantiomers, stereoisomers that **ARE** mirror images, like the left-handed and right-handed versions of amino acids.

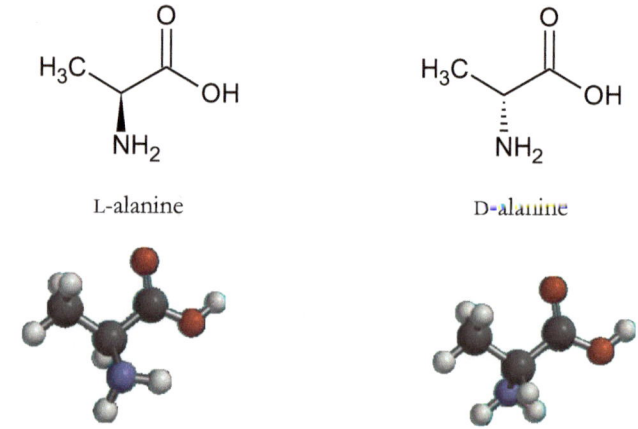

Molecular shapes are predictable from a very simple model, VSEPR theory (Valence Shell Electron-Pair Repulsion theory). This is described in your textbook; it works with another very simple model, Lewis structures. Lewis structures divide electrons into pairs that may be shared (bonds) or unshared ("lone pairs").

In VSEPR theory, each atom-to-atom connection or non-bonding "lone" pair of electrons represents an "electron group." Non-bonding "lone" pairs of electrons take up more room than shared electron pairs. Multiple bonds (such as in carbon dioxide, O=C=O) act each as a single group, though obviously a double or triple bond — with two or three times the number of electrons — takes up more space than a single bond.

Basic molecular shape can be predicted from the number of "electron groups" around a central atom:

- Two groups arrange themselves 180° apart, in a **linear** geometry.
- Three groups arrange themselves 120° apart, in a **trigonal planar** geometry.
- Four groups arrange themselves 109.5° apart, in a **tetrahedral** geometry.
- Five groups arrange themselves in a **trigonal bipyramid**, with three *equatorial* groups "trigonal planar" to each other (120° apart) and two *axial* groups "linear" to each other. The axial groups make a 90° angle with the equatorial groups.
- Six groups arrange themselves 90° apart, in an **octahedral** geometry.

Real molecular geometries are not usually perfect examples of this:

- While BF_3 is a perfectly trigonal planar molecule, formaldehyde ($O=CH_2$), while planar, has bond angles around carbon that deviate from the perfect 120°.
- While methane (CH_4) is a perfectly tetrahedral molecule, methylene chloride (CH_2Cl_2) has bond angles that deviate from the perfect 109.5°.
- Ammonia has four groups around nitrogen: three N-H bonds and one lone pair ($:NH_3$), so that it is considered "tetrahedral" by VSEPR. The H-N-H bond angles are compressed by the lone pair to about 107°. Because we don't normally include electrons in descriptions of molecular shape, we say that ammonia is "pyramidal" or "trigonal pyramidal."
- Water (H_2O) is also "tetrahedral" by VSEPR because there are *two* unshared pairs of electrons on the oxygen atom. Its H-O-H bond angle is compressed to about 105°. Because we don't normally include electrons in descriptions of molecular shape, we say that water is "bent."

Here are Lewis structures of the compounds discussed above (unshared electron pairs on fluorine and chlorine have been omitted for clarity):

$$\underset{F}{\overset{F}{\diagdown}}B\!-\!F \qquad \underset{H}{\overset{H}{\diagdown}}C\!=\!\ddot{\underset{..}{O}} \qquad H\!-\!\underset{H}{\overset{H}{\underset{|}{C}}}\!-\!H \qquad H\!-\!\underset{Cl}{\overset{H}{\underset{|}{C}}}\!-\!Cl \qquad H\!-\!\underset{H}{\overset{H}{\underset{|}{\ddot{N}}}}\!-\!H \qquad :\!\underset{H}{\overset{..}{O}}\!-\!H$$

Crystal structures

The macroscopic shapes and patterns of crystals are consequences of the way atoms are arranged within the crystal. Amorphous materials don't usually grow facets and other sparkly surface features, though they can have them introduced artificially (like cut glass).

Wikimedia

Here are two web pages that give (imperfect) models of the atomic structure of some crystalline substances:

- www.bluffton.edu/~bergerd/classes/NSC105/molecules4a.htm gives examples of salts, with various combinations of ions including salts of some polyatomic ions.
- www.bluffton.edu/~bergerd/classes/NSC105/molecules3.htm gives the structures of diamond and graphite, among other carbon-based structures.

Note that crystal growth in something like a crystal garden is the same sort of process by which other crystals (including diamond) can be grown. Normally, crystals are grown from a water solution by slowly allowing the water to evaporate, but diamonds are grown from carbon in the gas phase! (See the "For Example" section at the end of Chapter 12.)

Adhesives

Adhesives work by a couple of mechanisms.

The simplest adhesives (like Elmer's glue) work by penetrating the pores of a porous material, then drying. Intermolecular forces hold the glue together in a solid mass, and the "fingers" of formerly-liquid glue that have penetrated the two surfaces also hold *them* together. The same process is used to make non-stick Teflon stick to frying pans: the surface is sand-blasted to make it rough, and liquid Teflon then penetrates the roughened surface before drying. This is called "mechanical adhesion."

Sometimes glue is formulated so that it reacts with itself to form a solid that holds the parts together. The advantage of this is that the curing time can be much less than the drying time of an adhesive like white glue, and chemically-bonded glue is normally much stronger. However, shrinkage is much more of a problem with glues of this type. Examples include epoxy and cyanoacrylate ("super") glues.

Vocabulary

You should write your own definitions for these words, based on the textbook and this outline:

Constitutional isomers

Stereoisomers

How does VSEPR tell us that electron structure leads to molecular shape?

Trigonal planar

Tetrahedral

Octahedral

Density

Amorphous

Crystalline

Surface tension

Surfactants

Questions

1. Why is ice less dense than liquid water? For help on this, see www.edinformatics.com/interactive_molecules/ice.htm

2. How does surface tension result from intermolecular forces?

3. Explain how adhesives work.

4. What are the important differences between Elmer's Glue and super glue? How are they similar?

5. Explain how surfactants are used to allow oil and water to mix.

PART 1, CHAPTER 13: "WHEN MATTER HEATS UP"

All images in this chapter are in the public domain.

Let's start by defining some terms:

Enthalpy is energy that is released or absorbed as heat; it's what your text refers to as "energy." But "energy" is a somewhat different concept (see "free energy", below). It's like the way we use "oxygen" for both the element (O) and the molecule (O_2, which is sometimes called "dioxygen"). To be fair, this is very common; "energy" is pretty good shorthand for enthalpy, because often enough we find that only enthalpy is important to the net energy change of a process.

Entropy is what your text refers to as "entropy." There are some common misconceptions being promoted by Cobb & Fetterolf, but we'll deal with them below.

Free Energy is what your text refers to as "free energy." It's a combination of enthalpy and entropy. The reason I prefer to use the term *enthalpy* is because that way, I can refer to "free energy" as just "energy."

Enthalpy

Enthalpy is what your text refers to as "energy," at least within Chapter 13. It is the energy that is released or absorbed as heat, and can (mostly) be measured by a change in temperature or a change in phase.

Enthalpy can be modeled successfully as molecules in motion, using kinetic molecular theory; we saw this in the context of gases, liquids and solids, Chapter 10. But since molecules are not featureless spheres like atoms are, there are three components of molecular motion:

- *Translation*, motion from place to place. Any object, including a single atom, can have translational motion, and heat energy is usually modeled as just the energy of translation (kinetic energy).
- *Rotation* is motion that involves spinning and tumbling. A featureless sphere like a single atom cannot have this sort of motion: how could you tell? But molecules are not featureless spheres; they have non-spherical shapes, and so they can have rotational modes of motion.
- *Vibration* is motion that involves parts of a molecule stretching and bending, that is, moving with respect to other parts of the molecule. Again, single atoms cannot have this sort of motion.

These three are sometimes referred to as *energy modes* of a molecule. A molecule's total enthalpy of motion is the sum of the enthalpy it possesses in each of the three modes. A substance at *absolute zero* (see Chapter 10) is said to have zero enthalpy, but in point of fact there is always a minimum, non-zero enthalpy for any substance. This is called the *zero-point energy* and arises because molecules can never be completely motionless. Zero-point energy[*] and the motion associated with it are consequences of quantum mechanics, which is largely beyond the scope of this course.

Temperature, in the kinetic molecular theory, is a measure of the average kinetic energy (that is, *translational* energy) of a substance's particles. However, translational energy, and rotational and vibrational energy, are not the only types of enthalpy that a molecule can possess.

There is also chemical enthalpy, which comes from chemical bonds between atoms, and the enthalpy of attractions between molecules.

When water boils, its temperature rises to the boiling point *and then stays constant*. That is because, at the boiling point, water molecules now have enough kinetic energy to break free of their attraction to

---

[*] Contrary to a number of scams you will find if you look for them, zero-point energy cannot be converted into useful work. Work is extracted from the *difference* between a higher energy state and a lower energy state, but zero-point energy is the *lowest* energy state possible. Therefore no work can be obtained from it.

other water molecules. This breaking free absorbs energy, and so the temperature does not rise because no more heat can be put into molecular motion until all the intermolecular attractions are overcome.

COMMON MISCONCEPTION ALERT: H_2O remains H_2O when it boils. H_2O does *not* break down into hydrogen and oxygen atoms, and hydrogen and oxygen atoms do *not* fly apart separately!

Another way of looking at this is to say that temperature is a measure of *average* kinetic energy. But at the boiling point, all the molecules with high values of kinetic energy are breaking free and leaving the liquid. So their (higher) kinetic energy values are no longer part of the average, and the average kinetic energy *in the liquid phase* remains the same even as more heat energy is added.

Energy of a chemical system also changes when one type of bond is exchanged for another. For example, photosynthesis

$$6\ CO_2 + 6\ H_2O \rightarrow C_6H_{12}O_6 + 6\ O_2$$

is *endothermic*; the reverse reaction is *exothermic*. In photosynthesis, the bonds in carbon dioxide and water are changed into the bonds in glucose and dioxygen (O_2).

Definitions:

- **Exothermic**: the reaction releases energy as heat (enthalpy).
- **Endothermic**: The reaction absorbs energy as heat (enthalpy).

Bond breaking is the reverse of bond formation, and the enthalpy absorbed by breaking a bond is identical to the enthalpy released when forming that bond. If a particular type of bond takes more enthalpy to break, we say that it is "stronger" than a bond that it takes less enthalpy to break. But *all* bonds require *some* enthalpy to break.

COMMON MISCONCEPTION ALERT: Breaking chemical bonds *does not* release energy. It soaks it up. What *releases* energy is *forming* chemical bonds. Almost without exception, every chemical process involves some bonds being broken and other bonds being formed; what makes it exo- or endothermic is the net energy over the whole process. To wit:

- If the bonds formed in a reaction are, on average, **stronger** than the bonds that were broken, the reaction is *exothermic*

because it takes less enthalpy to *break* the weaker bonds than the enthalpy you get from *forming* the stronger bonds.
- If the bonds formed in a reaction are, on average, **weaker** than the bonds that were broken, the reaction is *endothermic* because it takes more enthalpy to *break* the stronger bonds than the enthalpy you get from *forming* the weaker bonds.

This applies to other sorts of interactions as well. For example,
- When lye (sodium hydroxide) dissolves in water, the sum of attractions between sodium ions and hydroxide ions (in solid lye) are weaker than the sum of the attractions between sodium ions and water molecules, and those between hydroxide ions and water molecules. Thus, dissolving lye in water is *exothermic*.
- When baking soda (sodium hydrogen carbonate or "sodium bicarbonate") dissolves in water, the sum of attractions between sodium ions and bicarbonate ions (in solid baking soda) are stronger than the sum of the attractions between sodium ions and water molecules, and those between bicarbonate ions and water molecules. Thus, dissolving baking soda in water is *endothermic*.

And yet baking soda still dissolves in water! If they take an input of enthalpy to happen, why do any endothermic processes happen at all?

Entropy

Your text perpetuates a common over-simplification, that entropy is disorder. **That's just not the case.**

"Disorder" is not even a good metaphor, because a particular "disordered" state is no less improbable, no lower in energy or higher in entropy, than a particular "ordered" state. And the word "disorder" doesn't get at the primary way that increasing entropy makes itself felt: **energy becomes more and more spread out, and less and less available to do useful work.**

The best definition of entropy* **is this:** "Entropy measures the spontaneous dispersal of energy: how much energy is spread out in a process, or how widely spread out it becomes—at a given temperature."

In other words, this is **not** necessarily an increase in entropy:

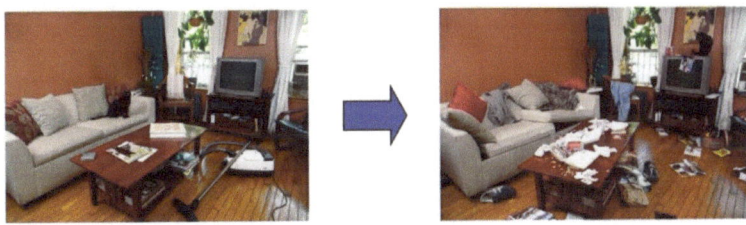

Entropy ≠ "Disorder"

But this may be a better metaphor:

Entropy = "Dissipation"

Free Energy

Whether a reaction is *spontaneous*, that is, whether it will happen without a continuous input of energy, is not based on its enthalpy, but on its overall *free energy*.

Free energy is the sum of two terms: the change in enthalpy, and the change in entropy. A release of enthalpy (*exothermic process*) is a favorable thing, and an increase of entropy is also a favorable thing, but… if there is an "intake" or "absorption" of enthalpy (an *endothermic process*) and yet there is enough of an increase in entropy to make up for that, the overall free energy change will be favorable, and the process will be spontaneous.

---

* Frank Lambert, "A Student's Approach to the Second Law and Entropy." entropysite.oxy.edu/students_approach.html#what (accessed May 24, 2011)

For example, when baking soda dissolves in water, the small *endothermic* enthalpy change is overcome by the large increase in entropy caused by breaking up the crystalline structure of baking soda and spreading sodium and bicarbonate ions through the water. Therefore baking soda spontaneously dissolves in water.

Scientists state this as $\Delta G = \Delta H - T\Delta S$, where G is free energy, H is enthalpy, T is temperature and S is entropy. Δ means "change in" so that "ΔG" means "change in free energy." Notice that a *negative* (less than zero) change in free energy means a release of free energy, and so is favorable. Likewise, a *negative* change in enthalpy is favorable, and a *positive* change in entropy is favorable (the subtraction term in the equation makes it negative).

You should be aware that, while a fair number of spontaneous processes are endothermic – notably dissolving certain ionic compounds in water – most spontaneous chemical reactions are exothermic. Very few useful industrial chemical processes are endothermic. A primary problem in industrial chemistry (or rather, chemical engineering) is getting rid of the heat produced by a desired reaction.

So why does photosynthesis happen if it's endothermic? Many or most metabolic processes, especially those that build big molecules out of small ones, are endothermic. Living things get away with that by literally dumping entropy into their environment. They make sure that their endothermic processes (like photosynthesis) include components that increase entropy somewhere else. *Otherwise they wouldn't work.* **The overall free energy change has to be favorable**, and if the *enthalpy* change is unfavorable, a lot of *entropy* has to be generated.

For example… refrigerators

Refrigerators are a simple example of this. They expend energy (running a compressor) to run an endothermic process (cooling something) and do it by dissipating energy (heat) into the environment, increasing the environment's entropy.

The Second Law of Thermodynamics* ("in the universe as a whole, entropy always stays the same or increases") guarantees, among other things, that the amount of waste heat dissipated into the environment by a refrigerator *will **always** be greater than the amount of heat absorbed by the cooling process.* Thus, you can't cool a room by just leaving the refrigerator door open, because the refrigerator is dumping its waste heat into the same room.

You *could* cool a room with a refrigerator *if* you put the heat-radiating coils on the other side of the wall. That's how air conditioners work: they use exactly the same cooling cycle as a refrigerator but put the cooling section and the heat-radiating section in different locations. Portable air conditioners dump their warm air through a hose that is supposed to be put out of an open window.

Vocabulary

You should write your own definitions for these words, based on the textbook and this outline:

Enthalpy

Entropy

Free energy

Exothermica

Endothermic

Spontaneous

---

* For an excellent discussion of entropy and the Second Law, go to 2ndlaw.oxy.edu (accessed 10 October 2013).

Questions

1. If you open the door of a running refrigerator in a perfectly-sealed room, will the room get warmer or colder? Explain.

2. People sometimes use the fact that you get heat by burning methane – changing it into carbon dioxide and water – to claim that energy is released by breaking chemical bonds. Explain why this is not true, and where the energy release comes from when you burn methane.

3. In terms of enthalpy, entropy and free energy, explain what conditions must be true for a process to be *spontaneous*.

4. Ammonium sulfate will dissolve in water, even though the process absorbs heat. Explain why this is true, or even possible.

5. Explain, on the atomic/molecular level, what happens when water boils.

6. What is the difference between an <u>irreversible</u> process and a <u>reversible</u> process? (Hint: it's not true that you can't get the irreversible process back to its starting point in *some* way.)

7. Charging your cell phone battery always makes the battery warm to the touch. Why is that?

PART 1, CHAPTER 14:
"A WHOLE NEW PHASE"

Zone refining, which is discussed in the Demonstration 14 section of *The Joy of Chemistry*, is an example of a **partitioning** equilibrium:

impurities dissolved in liquid ⇌ impurities dissolved in solid

Impurities prefer to be dissolved in a liquid rather than in a solid, and so if time is allowed for the equilibrium to adjust, impurities move out of the solid into the liquid phase, and thus move down a molten metal bar as it is slowly allowed to solidify from one end.

Partitioning is also involved in the video demonstration on Moodle, showing how neutral and ionized forms of indophenol prefer different layers. As indophenol molecules (which prefer to be in the non-polar phase) are changed into indophenolate anions (which prefer to be in the polar phase), they preferentially move into the phase they favor.

Phase changes are an example of *reversibility*, in which a process can go either forward or backward along the same pathway. In a reversible process, going backwards exactly reverses everything.

- If the color changed from red to blue, the color in the reverse process will change from blue to red.
- If the forward process involves going from a solid to a liquid, the reverse process will involve going from a liquid to a solid.
- *If the forward process is endothermic, the reverse process will be exothermic.*

Phase changes as reversible processes

Any material can (at least in principle, if not always in practice) undergo three types of reversible phase changes: between solid and liquid, between liquid and gas, and between solid and gas.

from…	endothermic forward process	exothermic reverse process	to…
solid	melting	freezing	liquid
liquid	evaporation / boiling	condensing (condensation)	gas
solid	subliming (sublimation)	deposition	gas

Once again, it cannot be over-emphasized that *if a process is endothermic, the reverse process will be exothermic.** This is why river water, used to condense steam at a power plant, gets hot: condensation is an *exothermic* process and steam gives up heat when it condenses. Meanwhile, evaporation of sweat cools your skin because going from the liquid phase to the gas phase is *endothermic*.

Phase diagrams: phase-change behavior at different temperatures and pressures

Read the discussion in your text about phase diagrams. Notice that this concept doesn't just apply to melting and freezing; it also applies to the solubility of solids, liquids and gases. As an example of phase behavior under varying pressure, water at low pressure will evaporate rapidly at room temperature, cooling itself enough to freeze. See the Moodle page showing "Water freezing as it boils."

I have also prepared a Moodle page for Chapter 14 with a series of videos showing the melting and refreezing of carbon dioxide. Carbon dioxide will not liquefy at a pressure lower than about 4 atmospheres (see the phase diagram on p. 208).

Read the chapter carefully. There is one point that I think requires a little more explanation, though.

---

* This is true even if the process is *not* reversible.

Eutectic mixtures and the eutectic point

When two or more things are mixed in random proportions, the melting behavior of the mixture undergoes a change in behavior from the pure substance. I will outline the behavior of a binary mixture (two substances); as you read this, refer to the diagram on page 214 in your textbook.

- When a pure, crystalline substance melts, you see a single melting temperature. There may be a mixture of solid and liquid until it all melts, but the temperature of the mixture stays at the same value until everything is melted.
- When a binary mixture of mostly A and some B melts, melting begins at a particular temperature, so that you see a mixture of solid and liquid. However, the temperature continues to rise as the solid continues to melt – very unlike the behavior of a pure substance.
  - The mixture begins by melting all the B, plus as much A as is able to mix with that amount of liquid B at that temperature.
  - As the temperature rises, more and more A can dissolve in the liquid B, until eventually all the A is melted. At this point, liquid A and B fully mix in the proportions of the mixture.
  - The proportion of A and B in the melt when the mixture first starts melting, is called the *eutectic composition*.
  - If a mixture of A and B is prepared that has the eutectic composition (a *eutectic mixture*), it all melts at the **same temperature**—just like a pure substance. This temperature is called the *eutectic temperature*.
  - On a graph of temperature vs. composition, the point at which the eutectic mixture melts is called the *eutectic point*.

Hopefully these bullet points will help you understand the diagrams of melting-point behavior. Solder, a mixture of various metals used for assembling and repairing electronic circuits, is a eutectic mixture. So is the mozzarella cheese used on pizza and lasagna.

Vocabulary

You should write your own definitions for these words, based on the textbook and this outline:

Equilibrium

Partitioning between two phases

Evaporation vs. boiling

Condensation

Sublimation (the verb is "to sublime" not "to sublimate")

Deposition

Metastable state

Suspension

Emulsion

Eutectic point

Eutectic composition

Eutectic mixture

Questions

1. A change from liquid to gas is always endothermic. Why, then, does water evaporate spontaneously?

2. Using the ideas of kinetic molecular theory, explain why melting is endothermic and freezing is exothermic.

3. Describe each of the following phase changes as endothermic or exothermic: condensation, deposition, evaporation, sublimation.

4. Describe what is meant by a *metastable state*. Give an example, and explain why it is metastable.

5. What is meant by a *eutectic mixture*? Explain why it is important in some applications.

6. Describe the changes that take place, including any changes in the external conditions, when moving from point A to point B on the phase diagram shown here. (The phase diagram is for carbon dioxide. Source: Wikimedia Commons.)

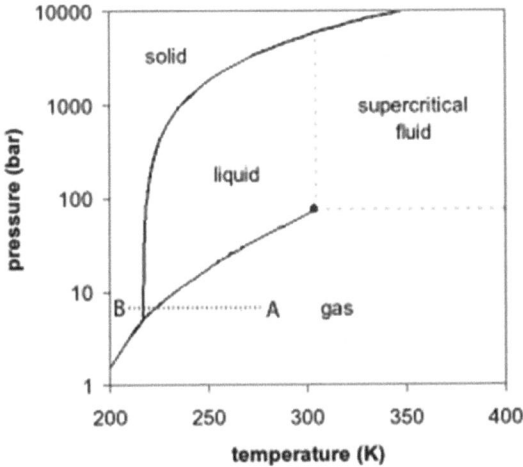

7. Why is it that liquids are not found in nature, except under certain conditions? What are those conditions? The diagram immediately above this question may be helpful in answering.

PART 1, CHAPTER 15: "EQUILIBRIUM — CHEMISTRY'S TWO-WAY STREET"

Equilibrium is a condition in a chemical reaction at which the ratio of reactants to products no longer changes. If that ratio is zero – that is, there are no reactants left – the reaction is said to have *gone to completion*.

Dynamic equilibrium

Equilibrium reactions do not stand still. Any particular set of atoms is constantly reacting, back and forth across the equilibrium.

In any chemical reaction that begins with reactants and no products, reactants are converted into products at first. But if the reaction is reversible, as soon as any product forms, it begins to be transformed back into reactants. Equilibrium has been reached when the ratio of products to reactants has reached a constant value. But remember that equilibrium does not mean that the chemical reaction has stopped: instead, reactants and products continue to be changed into each other at rates just right for the ratio of products to reactants to remain constant.

Equilibrium mixtures

In principle, *every* chemical reaction is an equilibrium, in which – once enough products have been formed – the reaction **appears to** shut down because reactants are being regenerated from products as fast as products are being formed from the reactants. An example is soda water, water with carbon dioxide dissolved in it. Dissolved carbon dioxide is in equilibrium with carbonic acid, H_2CO_3, which is why soda water has low pH. But forcing dissolved carbon dioxide out of solution will raise the pH of your soda water, because the remaining carbonic acid will continue to decompose into carbon dioxide even as the carbon dioxide is removed.

Try this experiment. Pour a small, clear glass about half-full of clear pop, and add some purple cabbage indicator (see the demonstration at the beginning of Chapter 4). The pop will turn pink, because its pH is about 3 or 4. Now set the (uncovered) glass in the refrigerator for a couple of days until the pop goes flat. What color is the indicator now? What is the pH of your flat pop?

Unbalanced equilibrium

In order to get an equilibrium reaction to go to completion, you need to selectively remove one of the products from the mixture. In Demonstration 15, carbonate ion – normally a miniscule part of a baking soda solution – is selectively removed by copper(II) ions, and so the baking soda is forced to react further with itself to form carbonate and carbonic acid. The carbonic acid product is in equilibrium with carbon dioxide and water, but because so much CO_2 is formed, the excess bubbles out of solution. The chemistry involved is discussed thoroughly on pp. 219-221. (You might remember that Demonstration 15 is a rerun of the first half of Demonstration 5, "Blue Blob, Black Ink.")

$$Cu^{2+} + 2\ HCO_3^- \rightarrow CuCO_3(\downarrow) + H_2CO_3 \rightarrow H_2O + CO_2(\uparrow)$$

This is an example of an *unbalanced equilibrium*. More and more of the bicarbonate ion changes into carbonate, while copper keeps the carbonate ion from changing back into bicarbonate, until either all the copper is gone or all the baking soda is used up.

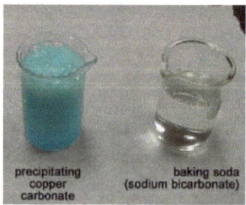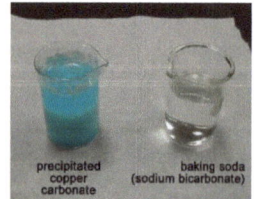

Images © by Daniel Berger

Notice how fizzy the blue mixture is in the middle photograph. This is because carbonic acid is being formed, and spontaneously decomposing to carbon dioxide and water. This second reaction is reversible, but its equilibrium strongly favors carbon dioxide. The CO_2 is supersaturated when it forms in such quantities, and it bubbles out of solution just as it does when a bottle of pop is opened.

Notice that this reaction is still reversible. If you were to provide something that had a stronger affinity for copper ions than carbonate does, the copper ions would be removed and baking soda would re-form.

Precipitation reactions are normally reversible, at least in principle, because there is **always** a **little** of the "insoluble" precipitate in solution. If this is chemically removed, the stuff is no longer precipitating as fast as it dissolves, and eventually the precipitate will disappear.

Frost-free freezers and freeze-drying (from the end of Chapter 14) are another example of setting up conditions for an *unbalanced equilibrium*. In this case, we have equilibrium between sublimation and deposition of ice: water vapor deposits at the same time as the ice sublimes, so that the amount of ice in a closed, ice-cold space does not change.

$$\text{water(solid)} \rightleftharpoons \text{water(vapor)}$$

But if dry air is circulated into the space, the water vapor is removed. While water continues to sublime, none of the vapor is permitted to deposit – and eventually all the ice in the space is gone.

$$\text{water(solid)} \rightarrow \text{water(vapor, removed)}$$

The last example (discussed in the "For example" section at the end of Chapter 15) is human respiration. The body creates situations in which there is more O_2 on one side of a membrane and more CO_2 on the other. The two gases diffuse through each membrane until equilibrium is reached – and that's how O_2 gets into our body and to our cells, and how CO_2 escapes.

Reversible and irreversible reactions

A reaction that does not go to completion, because it eventually attains equilibrium between products and reactants, is called a *reversible reaction*. In principle, **every** chemical reaction is reversible, but not every reaction forms an equilibrium mixture. In reversible reactions, the equilibrium ratio* (the amount of products divided by the amount of reactants) may be close to zero or very large, but forcing the reaction by removing one of the components – or supplying one of them in very large excess – can allow you to adjust the amount of the other components of the equilibrium to any desired values. These values must be consistent with the equilibrium ratio: the ratio of products to reactants never changes as long as the system is in equilibrium.

There are many reactions that are *irreversible*, because their reverse reactions are negligible under common conditions. The most familiar example is combustion. You don't see carbon dioxide and water changing into gasoline and oxygen under any conditions whatsoever.†

Energy, entropy and equilibrium

The composition of any equilibrium mixture is determined by the overall free energy of the process.

- A process with a free energy close to zero will have a nearly 1:1 ratio of products to reactants.
- A process with a free energy much more than zero (very positive) will have a very small ratio of products to reactants, nearly zero (almost no products present).
- A process with a free energy much less than zero (very negative) will have a very large ratio of products to reactants, approaching infinity (almost no reactants present).

Since free energy is composed of both enthalpy (based on the strength of chemical bonds and other interactions) and entropy (based on to what degree energy is spread out during the reaction), either or both of the two (enthalpy and entropy) can determine whether the free energy is favorable (negative) or unfavorable (positive).

In many common processes, entropy dominates, making the process spontaneous even if it is endothermic. Examples include dissolving ammonium sulfate in water (chemical cold packs) and the evaporation of sweat.

---

* Also called the "equilibrium constant."
† Exception: living things (plants and bacteria) can convert carbon dioxide and water into sugars and even into "gasoline."

Processes with both very large, favorable enthalpy and very large, favorable entropy tend to be classed as *irreversible*. The best example is combustion.

Vocabulary

You should write your own definitions for these words, based on the textbook and this outline:

Equilibrium

Balanced equilibrium

Unbalanced equilibrium

Equilibrium mixture

Reversible reaction

Irreversible reaction

Questions

1. An example of a reaction that is at equilibrium is shown below. What do we mean when we say that "equilibrium is dynamic"? You may explain in terms of the reaction shown.

$$NH_3 + H_2O \rightleftharpoons NH_4^+ + HO^-$$

2. "In a saturated solution, no more solid is able to dissolve." Explain how and why this statement is false, on the molecular level. In what way is the statement true, on the molecular level?

3. In terms of equilibrium, explain the difference between a reversible reaction and an irreversible reaction.

4. What is true of an *equilibrium mixture*?

5. How can equilibrium be "unbalanced"? What happens to the components of a mixture if their equilibrium is unbalanced?

6. In terms of equilibrium, explain why the temperature of boiling water does not rise above 100°C (212°F).

7. In the precipitation reaction between bicarbonate ions and copper ions to form solid cupper carbonate, where do the carbonate ions (CO_3^{2-}) come from? How are they formed?

   Cu^{2+} *(aqueous)* + HCO_3^- *(aqueous)* → $CuCO_3$*(solid)*   (not balanced)

8. Suppose that we mix equal amounts of water containing natural hydrogen (H_2O or HOH) – natural hydrogen consists almost entirely of the hydrogen-1 isotope – and water containing only the hydrogen-2 isotope, symbolized by "D" (D_2O or DOD). After a few days, all the water molecules in the mixture will have one atom of hydrogen-1 and one atom of hydrogen-2 (DOH). Explain this. What could be going on to make this happen?

Part 1, Chapter 16: "Colligative properties — strength in numbers"

The key points to note in this chapter are the ways that colligative properties are explained in terms of kinetic molecular theory, and in terms of dynamic equilibrium.

To sum up: after reading Chapter 16, you should be able to explain the four colligative properties (freezing-point depression, boiling-point elevation, vapor-pressure lowering and osmosis) in terms of

- a) Intermolecular interactions (or other processes on the molecular level)
- b) Dynamic equilibrium: how is equilibrium adjusting itself in the given situation? and
- c) Entropy. Be careful to translate the descriptions of Cobb & Fetterolf, which use the poor metaphor, "disorder," into a description that uses the much better metaphor, "dissipation" or "spreading-out."

Osmotic pressure

Osmosis is the process by which water will spontaneously flow through a semi-permeable membrane in order to bring solutions on either side of the membrane to an equivalent concentration: water will flow through the membrane from the more-concentrated side to the less-concentrated side, just as pure water poured into a glass of salt water will eventually spread itself evenly throughout the glass – diluting the salt water in the process.

As water passes through a semi-permeable membrane to dilute the more-concentrated side, the increased fluid volume puts pressure on the container on that side. This is a measurable physical pressure, like air pressure, exerted on the container on the more-concentrated side of a semi-permeable membrane. We can counteract this tendency by exerting pressure (as with a piston) on the more-concentrated solution; the pressure needed to counteract osmosis is called the *osmotic pressure* of a solution, and the solution **into which fluid will flow** during osmosis is said to have a higher osmotic pressure.

This definition means that osmotic pressure can be thought of as a "suction" effect: a solution with higher osmotic pressure will "suck" fluid through a membrane. It can also be thought of as the pressure exerted on its container by a more-concentrated solution as water flows into it.

Three definitions related to osmotic pressure are used to describe fluids used in medicine and other biochemical applications. This is because cell membranes are semi-permeable, and water will flow into or out of cells depending on the condition of the fluid surrounding the cells.

A fluid that has the same osmotic pressure as the normal fluids in living things is called *isotonic*. There is no net flow of water into or out of cells that are immersed in an isotonic fluid.

A fluid that has a higher osmotic pressure than the normal fluids in living things is called *hypertonic*. Cells immersed in a hypertonic fluid will shrivel up as water flows out of them. This is why salt is used for food preservation: its high osmotic pressure dehydrates bacteria.

A fluid that has a lower osmotic pressure than the normal fluids in living things is called *hypotonic*. Cells immersed in a hypotonic fluid will rupture: water flows into them until their cell membranes can't hold any more and burst, like blowing up a balloon.

So why is it that we can't preserve food by soaking it in distilled water? Well, for one thing, in contact with food the distilled water will rapidly pick up solutes and stop being pure water. The other reason is that decay bacteria are used to living in hypotonic fluids: they have cell walls to prevent rupture. But since those cell walls have to allow things to move through them, decay bacteria are vulnerable to hypertonic environments such as pickling brine.

Reverse osmosis

We can take advantage of osmotic pressure in order to force osmosis to go in the opposite direction. If we exert a strong pressure on the more-concentrated side of a semi-permeable membrane, we can force small molecules (notably water) through the membrane against their normal direction of flow. This is called *reverse osmosis*. Commercial reverse-osmosis water purifiers use a piston to exert pressure; this is true of both large-scale and small-scale (survival-kit) units. But it doesn't matter how the pressure is exerted.

In the "For example" section, which discusses kidney function, notice that one of the things kidneys do is apply pressure to the blood to squeeze out small molecules like water. The kidneys also use special molecules called *transporters* to keep from losing desirable small molecules that would otherwise be forced out in the reverse osmotic process.

Vocabulary

You should write your own definitions for these words, based on the textbook and this outline:

Colligative properties

Freezing-point depression

Boiling-point elevation

Vapor pressure

Vapor-pressure lowering

Osmosis

Semi-permeable membrane: give examples

Osmotic pressure

Reverse Osmosis

Questions

1. Explain osmotic pressure in terms of molecular interactions with other molecules and with a semi-permeable membrane.

2. Explain osmosis in terms of entropy.

3. Explain vapor-pressure lowering (or boiling-point elevation) in terms of intermolecular interactions.

4. Explain freezing-point depression in terms of entropy.

5. Why must energy be expended to get reverse osmosis to happen?
    a. Explain using concepts on the molecular scale.
    b. Explain in terms of entropy.

PART 1, CHAPTER 17: "CHEMICAL KINETICS — A VERITABLE EXPLOSION"

Reaction rates are affected by a number of things that are not microscopic (that is, they can be discussed without having to bring in molecules from the word 'go'). But the two most important are *concentration* and *temperature*.

Concentration. As seen in the demonstration, reaction rates increase with the concentration of reactants.* However, in a few reactions a necessary component of the mixture actually suppresses the reaction, so that rates *decrease* with reactant concentration. Please note that rates normally **do not** increase with concentration in a linear fashion. Often, they approach a maximum after which they no longer increase.†

Temperature. As seen in the demonstration, rates often increase with temperature. Again, this is not always a linear relationship, and sometimes increasing the temperature can *decrease* the rate.

Normally processes in which rate *decreases* with increasing temperature are exothermic processes which have an unfavorable entropy change; if you remember, $\Delta G = \Delta H - T\Delta S$, that is, the change in free energy is proportional to the change in enthalpy (heat of reaction) <u>and</u> to the temperature times the change in entropy.

---

* This is masked in the demonstration from the text, in which you use the same amount of bleach but different amounts of dye, because the concentration that controls how fast that reaction happens is the concentration of bleach.
† This is because, once concentration is high enough that reactants can find each other very quickly or instantly, increasing the concentration is not going to help them find each other any faster.

What this means is that if a process is exothermic (negative ΔH) but has an unfavorable entropy change (negative ΔS and hence positive $-T\Delta S$), increasing the temperature (T) enough will cause the unfavorable entropy term to dominate, and the reaction will shut down.

The very fact that the reaction is exothermic will raise the reaction temperature, and increase the effect of entropy!

Collision theory of reaction rates. This is based on the fact that two molecules have to collide in order to react. But they must also collide properly:

- They must collide in the correct orientation (for example, H_3O^+ must collide with the reactive portion of a base molecule in order to transfer H^+).

$$H_3O^+ + \underset{\text{acetate}}{\overset{O}{\underset{-O}{\overset{\|}{C}}-CH_3}} \longrightarrow H_2O + \underset{HO}{\overset{O}{\overset{\|}{C}}-CH_3}$$

$$H_3O^+ + H_3C-\overset{O}{\underset{O^-}{\overset{\|}{C}}} \longrightarrow \text{no reaction}$$

- They must collide with enough energy to force the reaction to happen.

The length of time it takes a reaction to happen can often be usefully expressed in terms of a concept called *half-life*, that is, the length of time it takes for half of what's there to be converted into products. Because the rate of reaction is dependent on concentration in an exponential fashion, it takes just as long to go from half of what you started with to a quarter of what you started with (half of a half), as it took to get from what you started with, to half of what you started with. So the half-life of a reaction is a clue to about how long it will take for the reaction to go to completion: typically about 8 or 10 half-lives will be enough for a complete reaction, if the reaction doesn't form an equilibrium mixture.

Activation energy

Your text calls this the *activation barrier*, because the energy of activation does in fact behave as a barrier to reaction. If it weren't for activation energies, we would all spontaneously slump into a bubbly pool of carbonated water and nitrogen gas, with traces of sulfuric acid and a few other oxides. Those are the implications of the fact that the chemical structures that make up our bodies are unstable with respect to both enthalpy (burning a human being releases heat!) and entropy.

In other words, activation barriers dam the second law[*] (of thermodynamics).

The example given at the beginning of Chapter 17 was the fact that diamonds are unstable with respect to graphite. If there were no activation energy for the conversion,[†] all our engagement rings and other sparkly diamond stuff would instantly become a shiny black powder. Fortunately, the activation barrier for this is quite high indeed. The half-life for conversion of diamond to graphite, under "normal" conditions of temperature and pressure, is as close to forever as makes no difference.

But suppose we change the conditions, by providing a pathway with a lower activation barrier? Then the conversion of diamond to graphite would go like a shot.

Many spontaneous chemical processes that have significant activation barriers can be speeded up by using something that does just that: **it provides a lower-energy reaction pathway** so that the rate of product formation increases. A substance that does this, while itself remaining unchanged, is called a *CATALYST*.

A technical note. Cobb & Fetterolf say that catalysts "lower the activation energy" for a reaction. This is not quite right. Instead, a catalyst **provides another pathway** that has a lower activation energy to get from reactants to products. See the difference?

The effect of a catalyst is to speed up the conversion of reactants to products. Often, reactions that won't happen in a useful way under

---

[*] Frank Lambert, "Shakespeare and Thermodynamics: Dam the Second Law!" shakespeare2ndlaw.oxy.edu/ (accessed May 24, 2011)

[†] Frank Lambert, "The Second Law of Thermodynamics (3)" secondlaw.oxy.edu/three.html (accessed May 24, 2011)

practical conditions can be made useful by the addition of a catalyst. An example that we have already discussed is the Haber-Bosch process for converting atmospheric nitrogen into ammonia: without the catalyst, the reaction would require much greater heat and pressure.

Vocabulary

You should write your own definitions for these words, based on the textbook and this outline:

Reaction rate

Activation energy or activation barrier

Half-life

Catalyst

Questions

1. Explain what an activation energy is. How does the activation energy indicate how fast a reaction will happen?

2. Explain how a catalyst works, in terms of activation energy.

3. Explain why increasing the concentration of one of the reactants will make a reaction happen more quickly.

4. Explain why decreasing the temperature at which a reaction happens will make that reaction happen more slowly.

5. Give two reasons why every collision between two reactant molecules does NOT lead to a reaction between those two molecules.

6. Explain why, if a reaction involves a reduction in entropy, increasing the reaction temperature will slow the reaction.

PART 1, CHAPTER 18:
"ELECTRONS AND PHOTONS — TURNING ON THE LIGHT"

The key concepts here are *photochemistry* and *electrochemistry*.

How photochemistry works

Light comes in chunks, called photons; shorter wavelength photons have more energy. Long-wavelength radiation like microwaves or infrared cannot do anything more than make molecules move around and bump into each other, while short wavelengths like ultraviolet light carry enough energy to make molecules change chemically even if they don't have other molecules around to react with.

Electrons occupy energy levels that are stair-stepped within an atom or a molecule. An electron must absorb a whole photon at once (or the equivalent in collision energy) in order to be *excited*, that is, promoted to a higher energy level. At that point it may do three things:
1. The electron may do chemistry by helping its molecule or atom to react.
2. The electron may jump back down (*decay*) to its lowest energy level in smaller steps, emitting a photon at each step.
    - *Phosphorescence* is time-delayed, so that an object glows for a time after it has been exposed to light.
    - *Fluorescence* has no time delay: when you shut off the stimulating light source, the emission stops. "Fluorescence" – though it's not called that – can also happen if things are heated very hot, as in fireworks or other flames.

- o Very hot objects glow red or even white, as the violent collisions of their atoms and molecules create excited electrons that then undergo decay.
3. The electron may jump all the way back down in one step, emitting a single, very energetic photon. This is rare but does occur.

Because electron energy levels are not continuous, but stair-stepped, an electron can't just absorb any amount of energy; it can only absorb an amount of energy corresponding to a jump of one, two, three, or some other whole number of energy levels. In the same way, an excited electron can only emit photons of a particular set of energies that correspond to dropping down a whole number of energy levels. For example, arsenic atoms will glow violet in a flame, copper atoms will glow green, and lithium atoms will glow red.

Wikimedia

Chemical reactions can be caused by light energies at or above the high-energy (blue) end of the visible light spectrum. This process is called *photochemistry*, and several examples are given in your textbook. Very-high-energy (short wavelength) photons, such as hard ultraviolet light or x rays, have enough energy to break chemical bonds directly. Weaker chemical bonds, such as those in bleach or hydrogen peroxide or bromine, can be broken by soft ultraviolet, such as that given off by a black light (Demonstration 18).

Several examples of photochemical reactions are given in Chapter 18.

How electrochemistry works

All electrochemical reactions are reduction/oxidation (redox) reactions: electricity must flow for electrochemistry to happen, and in at least part of the circuit, that means electrons have to move from one place to another. When electrons flow from one substance to another in a chemical reaction, that reaction is a redox reaction by definition.

There are two types of electrochemistry:
- *Electrolysis*, in which electricity is consumed. An *electrolytic cell* is an electrochemical setup in which electricity is consumed to

drive a redox reaction "uphill". An example is the chlor-alkali process that was discussed at the end of Chapter 7. Aluminum smelting is also an electrolytic process, as is recharging a battery.

- *Voltaic cells* or *galvanic cells* are electrochemical setups that generate electricity by using "downhill" redox reactions and drawing off the electrons that are traded to do work. The most common example is a flashlight battery, which is discussed in your text; but every type of battery acts as a voltaic cell when it is discharging (supplying electricity).

An electrochemical cell is connected to an external circuit by means of two *electrodes*. The *cathode* (positive) is the electrode to which electrons flow; the *anode* (negative) is the electrode that supplies electrons to the circuit.

Electrons, on their journey from anode to the cathode, carry current through the circuit outside the cell, and transfer energy from the cell to the load. But within the cell, current is not carried by electrons; instead, it is carried by ions of the *electrolyte*.

Electrolytes

For electric current to flow, you must have moving, electrically charged particles of some sort. Commonly, these particles are electrons – but what about a medium through which electrons cannot flow?

Water is an insulator because electrons cannot flow through it... but if you dissolve an ionic compound in the water, electrical current can flow because the ions can move through the water. A substance that conducts electricity when dissolved in water is called an *electrolyte*.

Most electrolytes are ionic compounds, M^+X^-. When they dissolve in water, the positive and negative ions are separated from each other by water molecules and can move freely.

Other electrolytes are acids, such as sulfuric acid. Sulfuric acid is a molecular compound H_2SO_4, but when dissolved it reacts with water to form hydronium and sulfate ions.

The way electrolytes allow current to flow in an electrochemical cell is by balancing charges. You see, if electrons flow from the anode to the cathode, the cathode quickly develops a negative charge that forbids any more electrons to come in, and the anode develops a positive charge that forbids any more electrons to leave.

Within the electrolyte, positive ions (*cations*) flow toward the cathode, balancing its negative charge; and negative ions (*anions*) flow toward the anode, balancing its positive charge. This continues (assuming an endless supply of electrons) until you have run out of ions in the electrolyte. Then the electrochemical cell shuts down.

Of course, you could also run out of electrons by using up either the anode material (the source of electrons, the reducing agent) or the cathode material (the electron sink, the oxidizing agent) before you run out of electrolyte. Battery design has to consider which of the three – anode, cathode or electrolyte – you want to use up first.

Corrosion

Corrosion is a redox process, and thus an electrochemical one. Electrons move from the metal to oxygen, forming a metal oxide. Water promotes corrosion partly because it can react directly with a metal (for example, iron) to generate a metal hydroxide and hydrogen gas; but it can also promote corrosion by allowing ions to move through it, carrying current that allows electrons to flow more easily from the metal to oxygen.

Batteries, in fact, are cases of *controlled* corrosion.

How batteries work

In a voltaic cell, the anode material and cathode material are separated by an electrolyte through which electrons cannot flow, but through which ions can move to balance the charges created by electron transfer. When an external circuit is connected to the battery, a chemical reaction occurs in which

- Electrons are transferred through the external circuit, from the anode to the cathode. The anode is oxidized and the cathode is reduced.
- Positive ions (*cations*) in the electrolyte move to the <u>cathode</u> to balance the negative charge created by reduction.

- Negative ions (*anions*) in the electrolyte move to the <u>anode</u> to balance the positive charge created by oxidation.

The anode and the cathode must not be in direct contact, or the electrons will be able to flow between them freely and you won't be able to get any work out of your battery. Instead, what you will have is a case of *uncontrolled* corrosion, such as what happens when copper and steel pipe fittings are screwed into each other by a poor plumber.

Image from doityourself.com

Non-rechargeable and rechargeable batteries

Non-rechargeable or "primary" batteries use electrochemical reactions that are not easily reversed. Once a primary battery is discharged, it cannot be recharged by forcing current through it; instead, it is likely to explode. Examples of this type of battery include not only standard flashlight and lantern batteries based on zinc and manganese dioxide, but also the lighter, longer-lasting lithium-manganese dioxide batteries sold for hearing aids and electronics.

Rechargeable batteries (storage or "secondary" batteries) can store electricity and release it when needed, because they use chemical reactions that can be easily reversed by simply forcing electric current through them. The most commonly-used batteries of this type are

- lead-acid batteries, with a lead anode, a lead dioxide cathode and a sulfuric acid electrolyte.
- nickel-cadmium batteries, with a cadmium anode and a nickel hydroxide cathode. The electrolyte is an alkali paste containing hydroxide ions.
- lithium-ion batteries use a carbon anode and a cobalt, iron or manganese ion cathode, with a lithium-ion electrolyte.

The key to a non-rechargeable battery is that it is able to be a *voltaic cell* but not an *electrolytic cell*.

The key to a storage battery is that it is able to function both as a *voltaic cell* (when discharging) and an *electrolytic cell* (when charging).

111

Corrosion protection: the sacrificial anode

The most obvious way to protect a metal from corrosion is to cover it with a barrier to water and oxygen. This is why, for example, bicycle frames are always painted.

But you can also protect a metal from corrosion by placing it in electrical contact with a more-easily-corroded metal. This can be as simple as coating iron with zinc ("galvanization"): even if the iron underneath is exposed to air and water, it will resist rusting as long as it can steal electrons from zinc.

Underground fuel tanks do this one better. There is no paint that will stand years of exposure to underground moisture and other soil chemistry. Instead, underground steel fuel tanks are electrically connected to a block of zinc, magnesium or aluminum in an easy-to-reach place that serves as an anode. So long as there is an active metal that is in electrical contact with the steel – and a way for anions from the soil to get to the anode to balance the negative charge it loses – the *sacrificial anode* will supply electrons to the iron, even as the iron loses electrons to the corrosive soil around it. When the sacrificial anode is depleted, it can simply be replaced.

Sacrificial anodes are also used to protect steel water heaters from corroding.

Vocabulary

You should write your own definitions for these words, based on the textbook and this outline:

Photochemistry

Excited electron

Electrochemistry

Electrolyte

Electrolytic cell

Voltaic cell or galvanic cell

Electrode

Cathode

Anode

Cation

Anion

Sacrificial anode

Questions

1. A chemical light stick and the fading of colors in the sun are both examples of photochemistry. Describe the similarities and differences between the two processes **as photochemical reactions**.

2. Explain the difference between fluorescence and phosphorescence, and give an example of each. How are they similar?

3. What three components are necessary for a battery?

4. Lead-acid batteries contain metallic lead plates and lead dioxide paste; discharging the battery consists of allowing the two to react in the presence of sulfuric acid to form lead sulfate. The chemical reaction is shown here:

$$Pb + PbO_2 + 2\ H_2SO_4 \rightarrow 2\ PbSO_4 + 2\ H_2O$$

Which is the cathode, metallic lead or lead dioxide? Which is the anode?

5. What is the essential difference between a primary battery and a secondary battery? Which is a true energy source?

6. Describe what a sacrificial anode does.

PART 2, CHAPTER 1: "SIMPLY ORGANIC"

All images in this chapter are © Daniel Berger

KEY POINT: in chemistry, "organic" simply means "built largely of carbon atoms."

Organic chemistry can be like architecture: it uses known structures and materials to build new things.

The most important reason that there are so many *different* organic compounds is carbon's unique ability to *concatenate*, that is, to form stable compounds that consist of rings and chains of carbon.

How unique is it? Well, this is addressed by one of Isaac Asimov's science essays, "The One and Only" from his collection *The Tragedy of the Moon* (Doubleday, 1973).* In a nutshell: there is *no other element* that can concatenate as easily as carbon can.

Because of the tetrahedral structure of carbon, and its ability to form stable bonds to itself, you can build intricate structures from it. These include steroids, proteins, and various pharmaceuticals.

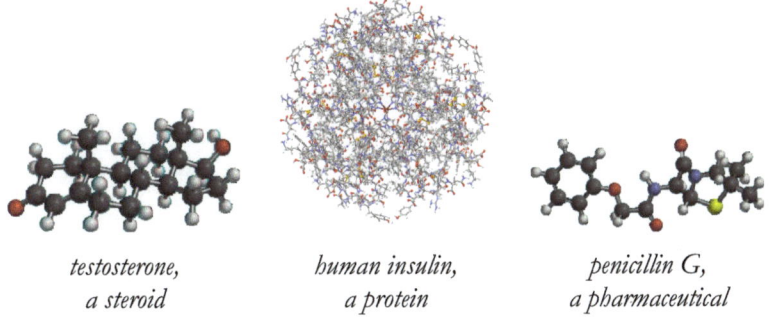

testosterone, a steroid   *human insulin, a protein*   *penicillin G, a pharmaceutical*

---

* Out of print, but the same arguments are made in this online article: www.madsci.org/posts/archives/2001-06/993247450.Ch.r.html (accessed 14 October 2013).

As we saw in the notes for Section 1, Chapters 7 and 12, the fact that carbon can form four bonds allows structures built of carbon to have *isomers*, that is, structures with the same formula (for example, C₈H₁₈) but different structural arrangements.

Below are the structures of the rings in Figure 2.1.5:

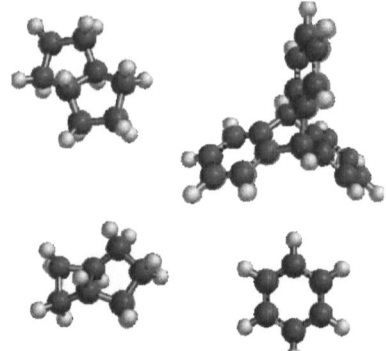

Isomers come in two types:

- Constitutional isomers have the same atoms arranged in different order (see the Chapter 7 notes).
- Stereoisomers have the same atoms arranged in the same order, but pointing different directions in space (see the Chapter 12 notes).

Organic compounds are modular

Organic compounds are often classified according to their *functional groups*. A functional group is a molecular subunit that has more-or-less the same chemical properties, no matter what else it is attached to. Some common functional groups are discussed in your text. Examples: *alcohols* all have the functional group "OH", *amines* all contain single-bonded nitrogen, and *carboxylic acids* all contain the group "COOH".

Being built of functional groups makes organic molecules modular: they can be thought of as a sum of their functional groups, and different overall properties can be had by tacking different functional groups onto the same framework, or by arranging the same functional groups around a different framework or in a different arrangement around the same framework. A single difference in functionality makes the difference between the male sex hormone testosterone and the pregnancy hormone progesterone, and exchanging alcohol groups for amine groups makes the difference between ethylene glycol, a toxic compound used for antifreeze, and ethylene diamine, used medically to treat heavy metal poisoning.

progesterone   *testosterone*

Note the convention here: "wedged" bonds are coming toward you, while "dashed" bonds are going away from you.

Biological molecules do this sort of thing *a lot* (see Part 2, Chapter 3). All proteins are constructed of the same 20 or so amino acids, in different arrangements and orders. Carbohydrates as different as glycogen and cellulose, or starch and table sugar, are constructed by arranging the same few simple sugars in different ways. And natural fats and oils all have essentially the same structure; they get different properties – usually different melting points – by using different modules in their otherwise identical structures.

Organic molecules can be chiral

Chiral objects are as close as the ends of your arms: as the text explains, your hands are not identical to their mirror images, and thus are *chiral*. Other examples of macroscopic, chiral objects include screws (which come in right- and left-handed forms), shoes, and your cellular phone (hold the keyboard up to a mirror: is it identical to its mirror image the way a paper clip is, or is it reversed?)

This can be the basis of a type of isomer. Two molecules that are mirror images of each other, but are not identical to each other, are called *enantiomers* (see also Chapter 12).

Biological systems are very good at using only one of a pair of enantiomers, and occasionally use the other of a pair for something different. Examples given in your text are carvone and limonene. (NOTE: your text gets the odor of limonene wrong. Only a single enantiomer is found in both oranges and lemons. The other, non-biological enantiomer smells like turpentine.)

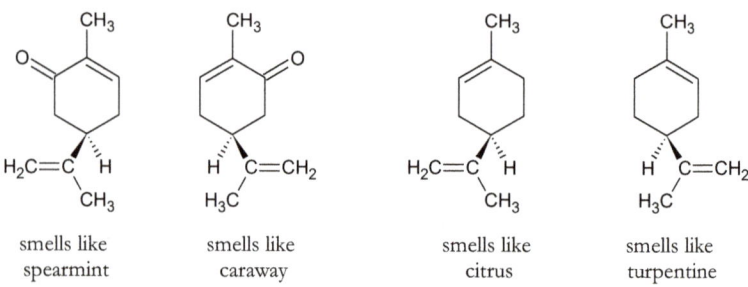

Since biological systems can usually only use one of a pair of enantiomers, the other one tends to be useless, and sometimes is actively harmful. The example given in your text is thalidomide: one enantiomer is a tranquilizer, but the other causes birth defects.

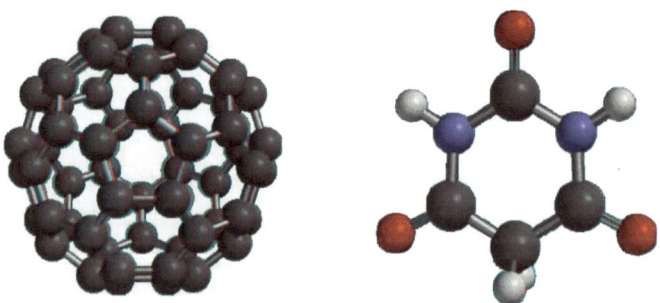

The enantiomers of thalidomide. The enantiomer on the left causes birth defects. Unfortunately, the body converts one enantiomer into the other. Thalidomide is now used to treat leprosy, but women patients are advised NOT to become pregnant.

Other structures mentioned at the end of the chapter: buckminsterfullerene (left) and barbituric acid (right).

Vocabulary

Write your own definitions for these terms, using the text and these notes.

Organic chemistry

Functional group

Isomer

Constitutional isomer

Stereoisomer

Chiral

Enantiomer

SECTION 2, CHAPTER 2: "CHEMISTRY ROCKS"

The authors (Cobb and Fetterolf) are talking through their hats when they talk about an "organic mindset." As noted in the supplement for Section 1, Chapter 7, carbon can combine in so many different ways that the vast majority (somewhere between 90% and 98%) of *naturally-occurring* chemical compounds are "organic" – they contain carbon in covalently-bound compounds.

Another problem with this chapter: at one point, in the discussion of the nitrogen and carbon cycles, Cobb and Fetterolf blur the chemical and colloquial definitions of "organic." Let's be clear:

- **Inorganic** carbon or nitrogen means that carbon or nitrogen is embedded in ionic compounds such as nitrates or carbonates, or in oxides such as CO_2, or (for nitrogen) not bound to carbon.
- **Organic** carbon or nitrogen means that the carbon or nitrogen atoms are embedded in covalently-bound compounds of carbon.
- The *colloquial* definition of "organic" is something produced by biological processes. This hasn't been the scientific definition for almost a century and a half.

Nevertheless it is true, in both the carbon and nitrogen cycles, that carbon and nitrogen is changed from inorganic forms to organic forms and back again.

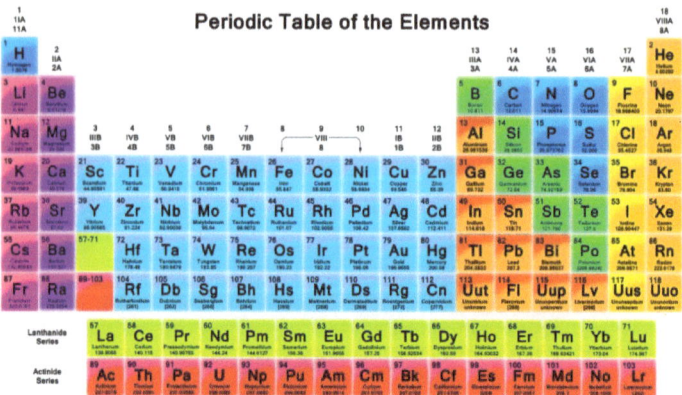

Periodic table image © Anne Marie Helmenstine

Element groupings

You should know the following divisions of the Periodic Table (see Section 1, Chapters 2 and 5):

- The green diagonal running from boron to polonium are called *metalloids* (or "semi-metals") and are commonly used in semiconductor materials. Elements to the left of these are metals (except for hydrogen); elements to the right (and including hydrogen) are non-metals.
- Families 1-2, 13-18 are the *main-group* elements (your text calls them the "representative elements" – a term that is not generally accepted).
  - Family 1 (excluding hydrogen, H) are the *alkali metals*.
  - Family 2 are the *alkaline earth metals*.
  - Family 17 are the *halogens*.
  - Family 18 are the *noble gases*.
  - Other families within the main group are often referred to by their first members, that is, "the carbon family" or "the oxygen family."
  - The classification "semiconductor elements" is invented by the authors. What is more, it is not entirely correct, when applied (as Cobb and Fetterolf do) to families 13 and 14. The common semiconductors gallium arsenide and cadmium telluride both contain elements that are *not* in the boron or carbon families.

- Families 3-12 are the *transition metals*.
  - More than most other elements, transition metals are known for their ability to move easily between several oxidation states – and the different colors associated with these oxidation states.
- The two rows below the main table fit into periods 6 and 7: between barium (Ba) and lutetium (Lu), and between radium (Ra) and lawrencium (Lr). They are called the *rare earth elements* and have almost identical chemical properties.

Transition metals and coordination chemistry

Many of the ligands used in natural and artificial coordination compounds are organic.

- Oxalic acid in radiator-cleaning fluid makes calcium and iron ions soluble in water.
- Chlorophyll and hemoglobin both use large organic structures to *chelate* (surround) metal atoms that do much of their work: magnesium in chlorophyll, iron in hemoglobin.
- Another organic coordination complex is Vitamin B-12, each molecule of which contains an atom of cobalt.
- Ethylene diamine (Section 2, Chapter 1 notes) is used to chelate heavy metals such as lead or mercury in the bloodstream, which prevents them from forming coordination complexes with important proteins, and thus deactivating the proteins.

The main-group elements ("representative elements")

Most of this part of the text is fine, except for the class of "semiconductor elements," which is an invention of the authors.

Main-group elements fall into distinct families with different properties: alkali metals, alkaline earth metals, boron and carbon families ("semiconductor elements"), nitrogen family ("pnictogens"), oxygen family ("chalcogens"), halogens, and noble gases.[*]

---

[*] Not "nobel" gases, **noble** gases.

Most main-group elements follow the octet rule; the exceptions tend to be when they are bound to fluorine or oxygen (see Section 1, Chapter 3: the oxidation-number rules). The noble gases don't have much chemistry simply because they don't need other elements to make their octets.

However, chemical stability has ***nothing*** to do with nuclear stability. For a discussion of the parts of the atomic nucleus and what holds them together, see Section 1, Chapter 2.

Radioactive elements and "radiochemistry"

Chemical arrangements that are unstable will tend to undergo reactions to form more stable arrangements.

In the same way, nuclear arrangements that are unstable will tend to undergo reactions, normally by *radioactive decay* in which a nucleus emits energetic radiation. This is what happens when the chemically-stable but radioactively-unstable element radon decays; if the radon atom happens to be in your lung, the energetic radiation will cause tissue damage (see "Ionizing radiation," below).

Radioactivity is a result of proton-neutron imbalances

As pointed out in Section 1, Chapter 2, neutrons are needed to supply the extra nuclear binding energy that holds positively-charged protons inside an atomic nucleus.

You can't pack in neutrons indefinitely because neutrons are unstable: they decay into protons and electrons. Neutrons that are not needed to hold the nucleus together tend to change into protons, emitting an energetic electron in the process. This is a form of radioactivity called *beta-minus emission* and changes the atomic number (and thus the chemical element) by +1.

But if you have too few neutrons, it's possible for a proton in the nucleus to capture an electron from just outside the nucleus, and change into a neutron. This is a form of radioactivity called *K-electron capture* and gives rise to a cascade of x-rays when electrons from higher in the atom's electron shells "fall" to the empty slot in the lowest electron shell. See Section 1, Chapter 18. K-electron capture changes the atomic number, and thus the chemical element, by −1.

A proton can also, more rarely, emit a positron, which has the same mass as an electron but an opposite charge; this is called *beta-plus emission* and is the basis of a medical imaging technique, *positron emission tomography*. Positron emission changes a proton into a neutron, changing the atomic number by −1.

As you move up in atomic number, you can't find a stable, balanced number of protons and neutrons any more: there are too many nucleons and you have to shed mass. After lead (atomic number 82), no element has any non-radioactive isotopes.* The most stable isotopes of each too-heavy element decay by emitting packets, called *alpha particles*, which contain two protons and two neutrons each. The new atom, with atomic number lowered by 2 from the parent and atomic mass lowered by 4, is usually radioactive because it has the wrong ratio of protons to neutrons.

Gamma radiation is a form of electromagnetic energy, like visible light or radio waves but much more powerful. It is emitted when a nucleus, newly formed by another radioactive decay process, "falls" into a more stable arrangement of nucleons. This is analogous to what happens in fireworks, when electrons emit visible light by falling into more stable arrangements (see Section 1, Chapter 18). For example, cesium-137 has too many neutrons, and undergoes *beta-minus decay* to form barium-137. The new barium nucleus – called barium-137m, for "metastable" – emits a gamma ray in order to shed excess energy and rearranges into a more stable state. Gamma-decay *does not* change the atomic number or the atomic mass.

Carbon dating

All radioactive atoms have a certain probability of decaying within a given time. If you have a large enough number of such atoms, some of them will be decaying at any given time, but not all of them decay at once.

The amount of radiation is proportional to the number of radioactive atoms present *and* how rapidly they tend to decay, and so radioactive decay (and the intensity of radiation) can be described in terms of *half-life*, the amount of time it takes for half of a sample of a particular radioactive material to decay (see the notes for Part 1, Chapter 17, for another discussion of "half-life"). For radioactive decay, the half-life is constant for any given isotope of a particular element.

- After one half-life, ½ of the original amount will be left.
- After two half-lives, ¼ (½ of ½) will be left.
- After three half-lives, 1/8 (half of a quarter) will be left.

---

* Bismuth, atomic number 83, comes close. Bismuth-209 is *almost* completely stable, with a half-life of 20 quintillion years (2×10^{19} years). P de Marcillac et al. *Nature* **422**, 876 (2003).

- ...
- After ten half-lives, the radioactivity is gone for all practical purposes: only 1 part in 1024 (2^{10}) of the original sample of radioactive material will remain. Ten half-lives is generally the length of time after which a dangerously-radioactive material is considered to be safe, that is, to have radiation levels no higher than "background" – the amount we get from natural cosmic, solar and earthly radiation.

Carbon-14, a radioactive isotope of carbon, is used to date materials that were once living. Carbon-14 is constantly being formed in the upper atmosphere, and is constantly being incorporated into living organisms, so that any living thing will have a more-or-less constant amount of carbon-14 within itself.

Nitrogen-14, the most common isotope of the most common element in our atmosphere, is converted to carbon-14 by certain types of cosmic-ray impact in the upper atmosphere. But carbon-14 has too many neutrons, and decays back into nitrogen-14 via beta-minus emission, with a half-life of about 5,700 years.

Carbon-14 is chemically identical to any other carbon isotope, and immediately reacts with oxygen to form carbon-14 dioxide. Since carbon dioxide is heavier than air, the newly formed $^{14}CO_2$ sinks into the lower atmosphere. Radioactive carbon-14 dioxide is chemically identical to normal carbon dioxide, and is included in the sugars and starches made from CO_2 by plants, which are in turn eaten by animals. Some of the radioactive carbon-14 is incorporated into body parts, such as stems or bones, in plants and animals. The carbon-14 in a living organism is steadily undergoing radioactive decay, but the organism is also taking in carbon-14 at a steady rate, and so a living organism attains a *steady-state concentration* of carbon-14 that does not change much over its lifetime.

As soon as an organism dies, it stops taking in carbon-14. But the carbon-14 in its tissues continues to decay at a constant rate. Comparing the amount of carbon-14 in a sample of dead tissue (for example, a bone or a piece of wood) with a similar sample from a living or newly-dead organism will give you a good idea of how old the dead tissue sample is, typically within a few thousand to a few tens of thousands of years. This estimate can be refined by comparison with similar samples whose age is known from other sources – tree rings, for example, or dated tombs.

Because the half-life of carbon-14 is 5,700 years, samples older than about 100,000 years must be dated using other methods, including radioactive isotopes such as potassium-40 or uranium-238.

Ionizing radiation

Smoke detectors use a small amount of a synthetic radioactive isotope, americium-241, in their detectors. Americium-241 emits alpha-particles, which knock electrons off of gas molecules in the air inside the detector. The resulting ions are detected by the tiny electrical current they produce. When smoke gets inside a smoke detector, it latches onto the ions and prevents this current from flowing. The disruption in the normal current level is what causes the smoke detector to go off.

All energetic radiation, including alpha-, beta- and gamma-radiation from radioactive materials as well as x-rays and cosmic rays, can knock electrons out of neutral atoms and molecules, generating ions. Such radiation is therefore called *ionizing radiation*.

By knocking electrons out, ionizing radiation disrupts chemical bonds – both in atmospheric gases and in living tissues. This is why gamma rays and x-rays can be used to kill bacteria: they break enough of the bonds in vital enzymes and nucleic acids that the bacteria cannot live. Radiation is used to sterilize heat-sensitive surgical supplies, like latex gloves, and also to kill bacteria in food. Milk that is sold in cartons at room temperature (you see this mostly in Europe) has been irradiated with either gamma rays or x-rays.

You should note the following:

Irradiated food is *not* radioactive. X-rays and gamma rays are not capable of causing the nuclear reactions that make things radioactive.

Cancer is caused by the disruption of cellular chemistry; only ionizing radiation and ultraviolet light are able to do that. Chemical reactions using ultraviolet radiation were demonstrated in Part 1, Chapter 18. Ultraviolet light, unlike x-rays or gamma rays, cannot penetrate your skin, so it cannot cause cancers other than skin cancer.

Cellular telephones and other sources of microwave energy **cannot cause cancer**, because microwaves are not energetic enough to cause chemical changes of any kind.

Vocabulary

You should write your own definitions for these words, based on the textbook and this outline:

Organic, organic carbon

Inorganic, inorganic carbon

Metal

Non-metal

Alkali metals

Alkaline earth metals

Halogens

Noble gases

Transition metals

Rare earth elements

Coordination complex

Chelate, chelation

Main-group elements

Radiochemistry

Radioactive decay

Ionizing radiation

Half-life

Carbon dating

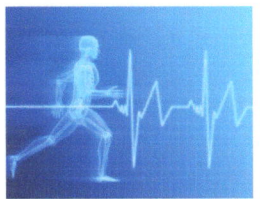

SECTION 2, CHAPTER 3:
"THE BODY OF CHEMISTRY MEETS THE CHEMISTRY OF THE BODY"

Images in this chapter © Daniel Berger

There's not a whole lot to add to the discussion in Cobb & Fetterolf. So here are the main points, with some extra images.

Polymers

Any chemical structure built of a number of modular subunits is called a *polymer*. The subunits from which polymers are constructed are called *monomers*. Artificial polymers tend to have just one or two types of monomer. The structures of three different, artificial polymers are shown below:

Polystyrene, *used for drinking cups and packing material, is made of a single monomer, styrene.*

Kevlar, *used for applications requiring great strength and low weight, is made of two monomers, phenylene diamine and terephthalic acid.*

129

PETA, *poly(ethylene terephthalate), used for applications such as carpeting and milk jugs, is made of two monomers: ethylene glycol and terephthalic acid.*

...-O-C(=O)-⟨⟩-C(=O)-O-CH₂CH₂-O-C(=O)-⟨⟩-C(=O)-O-CH₂CH₂-O-C(=O)-⟨⟩-C(=O)-O-CH₂CH₂-...

with the ethylene glycol monomer and terephthalic acid monomer labeled.

Biopolymers

Living things are largely constructed of polymeric, or at least modular, materials: carbohydrates, lipids, proteins and nucleic acids.

Carbohydrates

Carbohydrates are the class of molecules built of sugars, and have the general formula $C_n(H_2O)_n$, where n is an integer. Carbohydrates range from simple mono- and disaccharides to large polysaccharides. Small carbohydrates are used to fuel living things, while polysaccharides are used for structure* and energy storage in both animals and plants. Most natural polysaccharides are built of glucose.

A few of the common monosaccharides

glucose, mannose, galactose, fructose, ribose

---

* We often think of cellulose as the only structural carbohydrate; it's used by plants for their cell walls, and to stiffen stems and leaves. But arthropods such as insects and crabs use a modified carbohydrate, chitin, to build their shells.

Three common disaccharides: table sugar, milk sugar and malt sugar

sucrose (glucose + fructose)

lactose (galactose + glucose)

maltose (glucose + glucose)

Cellulose, an indigestible polysaccharide *(6 glucose units shown)*

Starch, a digestible polysaccharide *(6 glucose units shown)*

Lipids: steroids and fats

Lipids are the class of biomolecules that are insoluble in water. There are two types of lipids: glycerides, which are built of highly-flexible fatty acids and glycerine, and steroids, which are based on a rigid four-ring hydrocarbon structure.

Triglycerides

Fats and oils* are called *triglycerides* because a single fat molecule is made up of three fatty acids attached to the same molecule of glycerine. A typical triglyceride is shown below, showing three different fatty acids; normally, fats and oils are constructed from a mixture of fatty acids, and the fatty acids are attached to each glycerine backbone in no particular order.

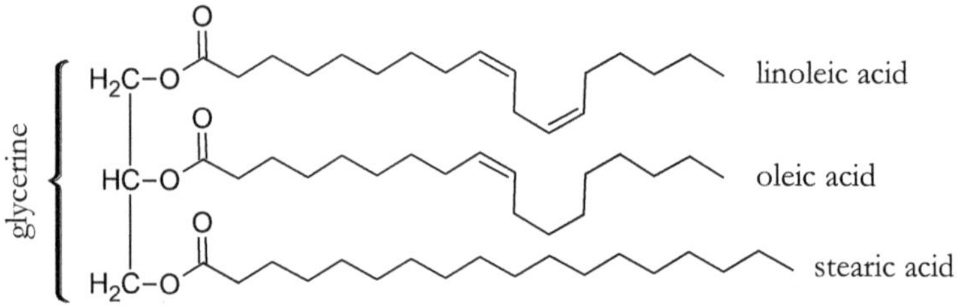

glycerine {
- H_2C-O—(C=O)—linoleic acid
- $HC-O$—(C=O)—oleic acid
- H_2C-O—(C=O)—stearic acid
}

Triglycerides are used not only for energy storage but for lubricating joints. Saturated fatty acids, such as stearic acid, melt at a higher temperature than unsaturated fatty acids, such as oleic acid or linoleic acid. Animals that live in temperate climates take advantage of this, changing the mix of fatty acids in their lipids as the weather changes, so that their joint lubricants always have about the same consistency.

Phospholipids

Living organisms also use lipids for their cell membranes, the barriers between cells and the outside world. Typically, *phospholipids* are used for this.

A typical phospholipid is a *diglyceride* (that is, glycerine with only two fatty acids attached) bound to a phosphate that performs some useful chemical function. For example, phosphitadylcholine has choline attached to its phosphate.

---

* The difference between a "fat" and an "oil" is that fats come from animals, while most oils come from plants (whale oil is an exception, but it's rendered from whale fat or "blubber"). There's also a difference in melting point: animal fats are solids at room temperature, while most plant oils are liquids. The most well-known exceptions are the solid plant oils, coconut oil and cocoa butter.

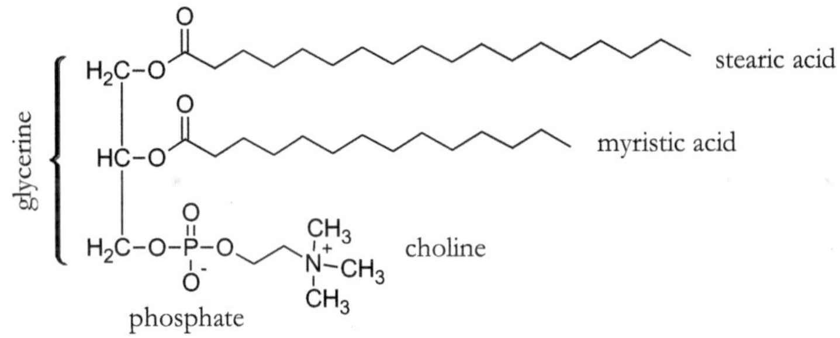

Other lipid types

Lipoproteins are protein-lipid complexes used for lipid transport in the bloodstream. The protein, being water-soluble, allows its attached lipid molecule(s) to move easily through blood, which is water-based.

Steroids are rigid four-ring molecules, such as cholesterol or the estrogens. In vertebrates, cholesterol is used in cell membranes, while most other vertebrate steroids are used as hormones, to regulate body chemistry and activity. Cholesterol is shown below, along with two of the many steroid hormones.

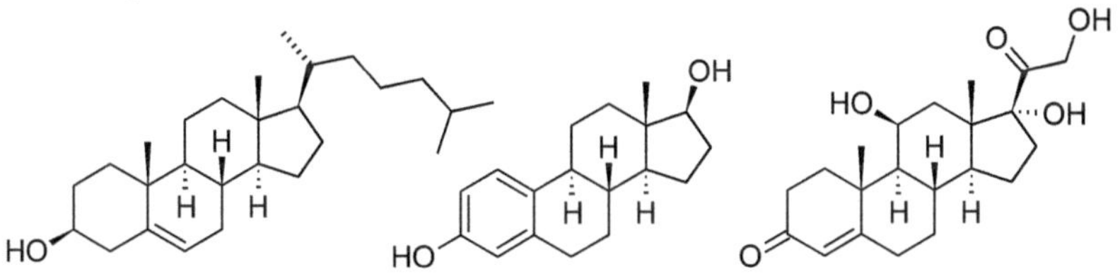

cholesterol estradiol cortisol

Proteins

Proteins are the workhorses: they are used for structural materials like keratin and collagen, and for nutrient transport, but as enzymes they also perform the everyday chemistry that keeps life going.

An amino acid is a compound that contains both an amine functional group and a carboxylic acid functional group. Biological amino acids have the structure shown at right, where "R" is a side chain. The identity of the side chain defines the amino acid.

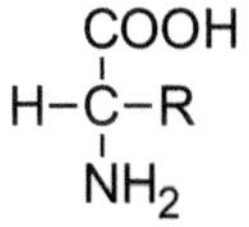

Proteins are polymers of amino acids; there are 20 amino acids that are used by vertebrates. By combining amino acids in different, specified sequences to make different proteins, a wide variety of tasks can be performed. Protein functions range from structure (e.g. collagen) to nutrient transport (e.g. hemoglobin) to detailed control of body chemistry by enzymes.

Amino acids are chiral (see Section 1, Chapter 12 notes and Section 2, Chapter 1), and the body only uses one stereoisomer of each amino acid. This means that each protein will have a specific three-dimensional structure, based not only on the detailed shapes of its constituent amino acids, but the way those amino acids interact with water and with other amino acids in the same protein.

A 10-amino-acid short section of human albumin, which is used to maintain fluid balance in the body and as a transport agent in the bloodstream. Albumin contains 585 amino acids. Notice the repeated unit in the backbone.

Nucleic acids

Nucleic acids are the materials that tell proteins what to do, by controlling how and when they are assembled. They, too, are polymers, but their monomers – called *nucleotides* – are more complicated than the monomers used for carbohydrates or proteins.

Nucleotides are, themselves, modular: each nucleotide consists of a sugar, a phosphate, and a base. The sugar and the phosphate are the same for every nucleotide, but the bases are different. There are four nucleic acid bases: adenine, cytosine, guanine and thymine/uracil.

adenine   cytosine   guanine   thymine

The two different nucleic acids use the same bases, except for one. DNA uses **thymine**, and RNA uses **uracil**; they differ only in a single methyl group.

There are two kinds of nucleic acid, which differ primarily in the sugar used (one of the bases has a minor difference as well). Nucleic acid that uses ribose is called *ribonucleic acid* (RNA), and the nucleic acid that uses deoxyribose – missing an oxygen from its #2 carbon atom – is called *deoxyribonucleic acid* (DNA).

The genetic code

Genetic information is stored in DNA and transcribed by RNA. DNA gives itself extra stability as a storage medium, as well as redundancy, by forming hydrogen bonds (Part 1, Chapter 8) between complementary base pairs on paired strands. These strands, because of the three-dimensional shape of their monomers, twist themselves into a double helix, a right-handed screw form. DNA is what our *genetic code* is built of.

RNA is used to read the code and transform it into proteins, using the manufacturing machinery of the cell.

The genetic code in DNA is written with four letters (the four bases), using words of three letters each. Each word stands for a different one of the 20 amino acids used in vertebrate biology. These words are transcribed by unzipping a portion of the double helix and building a strand of RNA, using the DNA as template. This *messenger RNA* then emerges from the nucleus and is, in its turn, used as a template to build a protein using three-letter words of *transfer RNA*. The transfer RNA words each grab their appropriate amino acid and are lined up as dictated by the strand of messenger RNA. When the protein has been completed, the transfer RNA falls away from it and is reused.

An RNA tetranucleotide, showing the four RNA bases

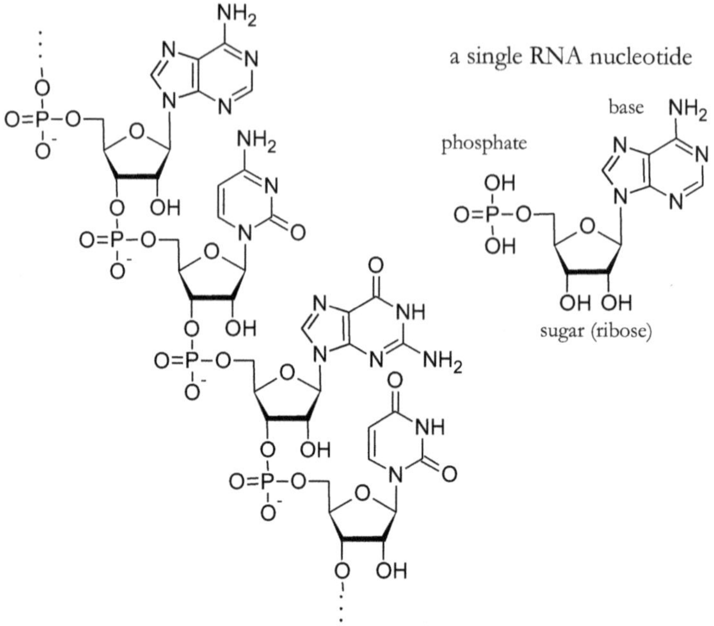

a single RNA nucleotide

A DNA tetranucleotide, showing the 4 DNA bases.

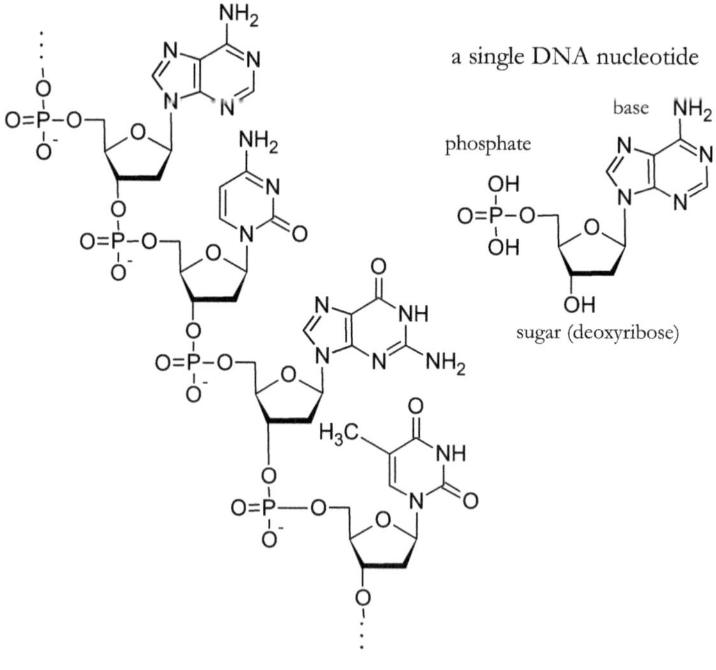

a single DNA nucleotide

Hydrogen bonding (dotted lines) **between DNA bases, and the double helix**

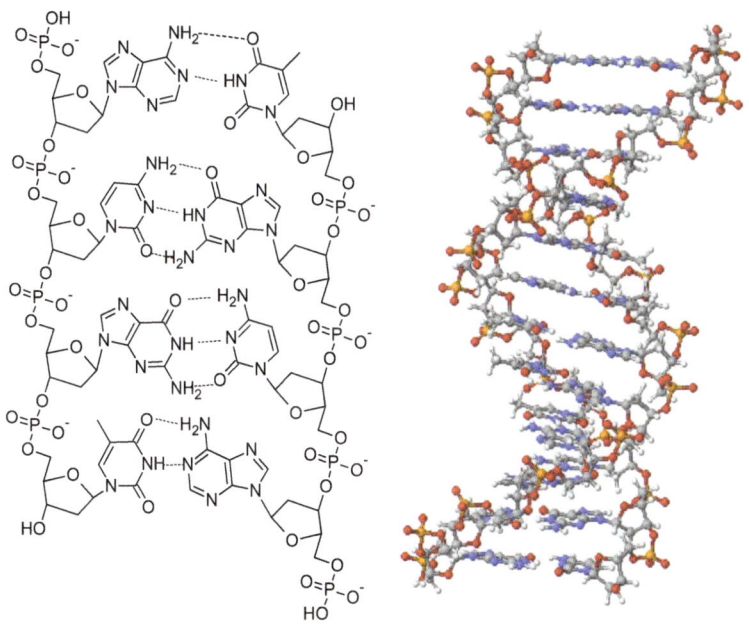

Vocabulary

You should write your own definitions for these words, based on the textbook and this outline:

Polymer

Monomer

Carbohydrate

Lipid

Protein

Nucleic acid

Sugar

Monosaccharide

Polysaccharide

Amino acid

Enzyme

Nucleotide

Nucleic acid base

Double helix

DNA (deoxyribonucleic acid)

RNA (ribonucleic acid)

SECTION 2, CHAPTER 4:
"CHEMIST AS ANALYST"

"What is it, and how much of it is there?"

All chemists must do analysis as part of their work. But an analytical chemist's work is to refine methods of analysis, and invent new ones. The discussion in the chapter is pretty good, but I want to expand on a few concepts.

Precision vs. accuracy.

Precision is a measure of how "tight" an instrument's measurements are, that is, how close repeated measurements are to each other. On the other hand, *accuracy* is a measure of how close your instrument's measurements approach the true value. Precision and accuracy are not necessarily related. The analogy that's often used is target shooting:

This target shows high precision, but low accuracy.   *This target shows low precision, but high accuracy.*

Wikimedia commons

Precision is affected by a number of factors, including *instrument noise levels*. Noise is simply the random signal produced by unpredictable environmental factors. Normally we want to minimize noise in any way that we can, but there are also techniques for *signal*

139

averaging that cancel noise and enhance signal by combining a number of individual measurements under identical conditions.

Confidence limits are expressions of the accuracy of a measurement. Confidence limits define a range within which you are confident the true value of a measurement lies. A confidence limit states how *accurate* you think your measurement is. If your confidence limits are close together, you are saying that your instrument is both accurate and precise.

Significant figures are an oversimplification of *confidence limits*. Normally you express your confidence in a measurement separately from its value (for example, 1.234 ± 0.123: 1.234 is the value, 0.123 is the confidence limit), so that the observer knows not only how much detail a measurement gives but also your estimate of how accurate (or precise) the number is.

But occasionally we might want to report just a rough value, with no confidence limit. In that instance, we would take the original number (1.234) and round it to the largest decimal place of our uncertainty (0.123). In this instance we would round our measurement to 1.2 and say that we know our value to "two significant figures." But doing this *destroys information* in our result, by lopping off useful digits that could be used in future measurements. Because rounding destroys information, it is never used in careful scientific measurements, but only in rough-and-ready reports. Any record that will be used for further work will always report a value, using every digit the instrument gives you, along with a statement of the confidence limits. You always record 1.234 ± 0.123, never just 1.2.

Analytical techniques: separation and identification

There are many analytical techniques in current use, but they all fall into two categories.

- **Separation** techniques are used to isolate the individual components of a mixture.
- **Identification** techniques are used to identify a particular substance.

Chromatography is a separation technique. The components of a mixture can be separated by *adsorbing* the mixture onto a *stationary phase* and allowing a *mobile phase* to flow past it.

The different components of the mixture divide their time between the mobile phase and the stationary phase, so that they spend part of their time moving and part sitting still. Each different component will spend different amounts of time in the mobile phase, so that eventually the mixture's components will be separated into *bands*. Each band can then be identified to find out what it is, and how much of it there is.

Some chromatographic techniques (such as gas chromatography or high-pressure liquid chromatography) hold their stationary phases confined inside a *column*, a long tube through which the mobile phase flows. This gives more precision than the paper chromatography demonstration discussed in your text, because there is no possibility of the mobile phase spreading sideways. (With paper chromatography we get the same effect by using thin strips of paper.)

Spectroscopy is an identification technique. The name is given to any technique that produces a *spectrum* (plural is *spectra*), defined in your text. The discussion in your text is adequate.

Spectra can be used to identify single substances in two ways:

By comparison. A spectrum is compared to a database of known spectra in order to identify a substance. This is by far the most common way of doing things in analytical laboratories.

By analysis. The features of some types of spectra (notably infrared and radio-wave spectra, but also mass and ultraviolet spectra) are associated with known structural features – **functional groups.** Spectral identification of the functional groups and how they fit together can reveal a chemical structure, just as fitting the pieces of a jigsaw puzzle together reveals a picture. Because the confidence level for this way of doing things is never more than about 90%, analytical laboratories prefer to do spectral identifications by comparison with known spectra.

Spectroscopic techniques are sometimes used as detectors for separation techniques. The most notable instance of this is the combination of gas or high-performance liquid chromatography with mass spectroscopy, allowing a chemist to simultaneously identify a compound and determine how much of it is present in a mixture.

Vocabulary

You should write your own definitions for these words, based on the textbook and this outline:

Analytical chemistry

Chemical analysis

Precision

Accuracy

Confidence limits

Noise

Signal

Why "significant figures" are less than ideal

Separation

Identification

Chromatography

Stationary phase

Mobile phase

Spectroscopy

Spectrum

Forensic science

Answers to Selected Problems and Questions

Chapter 1

1. What constitutes a chemical change? How is it different from a physical change? How would you tell?

   In a chemical change, at least one substance changes its chemical identity – that is, it changes into a different compound. In a physical change, chemical identity does not change.

   Being able to tell is a different matter, and more difficult: you have to establish the chemical identity or identities both before and after.

2. If something with one name changes into something with another name, is that always a chemical change? Can you think of an example when it is not?

   Simple example: water changing to ice. Both have the chemical formula H_2O, and the chemical identity "water."

3. If a substance is not detected, does that mean it is not there? Why would we say, for example, that "there are no pesticides in this organic produce"? Is that a reasonable thing to say?

   It is impossible to prove a negative; a physical measurement can only say that "nothing was detected," not that "nothing was there."

   If you know the detection limits of your equipment, you can exclude values higher than that limit. But you can't say that the value is zero.

Chapter 2

1. How many valence electrons does each of the following elements have? The atomic number and atomic symbol are given for each.
   a. Sodium (Na, 11)

   Sodium is in Family 1, and so has 1 valence electron.

   d. Oxygen (O, 8)

   Oxygen is in Family 16, and so has 6 valence electrons ("drop the leading 1").

   g. Fluorine (F, 9)

   Fluorine is in Family 17, and so has 7 valence electrons ("drop the leading 1").

2. Using the octet rule, predict the common valence of each element listed below, that is, the number of chemical bonds it is likely to form. The atomic symbol and atomic number are given for each.
   a. Sodium (Na, 11)

   The nearest member of the noble gases (Family 18) to sodium is neon, element #10. To get from 11 to 10 you have to drop one, so sodium will use one electron to form one bond.

   e. Oxygen (O, 8)

   The nearest member of the noble gases (Family 18) to oxygen is neon, element #10. To get from 8 to 10 you have to add two, and so oxygen needs to form two bonds to get those two electrons.

   Alternatively, oxygen has 6 valence electrons, and wants 8, so it needs two more electrons. It must therefore form two bonds.

   h. Fluorine (F, 9)

   The nearest member of the noble gases (Family 18) to fluorine is neon, element #10. To get from 9 to 10 you have to add one, and so fluorine needs to form one bonds to get that one electron.

   Alternatively, fluorine has 7 valence electrons, and wants 8, so it needs one more electron. It must therefore form one bond.

Chapter 3

1. What is the oxidation number for tin in stannous fluoride, SnF_2?

    By Rule 1, F must have an oxidation number of –1. Two fluorines add up to –2. To balance that, tin must have an oxidation number of +2.

3. What is the oxidation number for iron in rust, Fe_2O_3?

    By Rule 2, O must have an oxidation number of –2. Three oxygens add to –6. To balance that, the two iron atoms must add to +6, and so iron must have an oxidation number of +3.

Chapter 4

2. When sulfuric acid is dissolved in pure acetic acid (the acid in vinegar), the following reaction takes place:

    $$H_2SO_4 + HC_2H_3O_2 \longrightarrow HSO_4^- + H_2C_2H_3O_2^+$$

    sulfuric acid   acetic acid

    In this reaction, which is the acid, and which is the base?

    Sulfuric acid loses a positive hydrogen (H^+), and acetic acid gains a positive hydrogen. Losing H^+ means that sulfuric acid is acting as the acid, and gaining H^+ means that acetic acid is acting as the base.

4. Why doesn't baking powder need vinegar or sour cream to make batter rise, but baking soda does?

    Baking powder contains baking soda as well as an acid that is released by heat during baking; the reaction of this acid with the baking soda releases carbon dioxide and results in rising.

    However, baking soda doesn't have any acid of its own. Therefore it works better if a mild acid (such as sour milk) is included in the batter. The disadvantage is that you have to use the batter right away, or the baking soda will be used up before you can bake it.

Chapter 5

4. Something that is "insoluble" will still dissolve. Explain this.

   "Insoluble" means that the amount of a substance that dissolves is "essentially zero." But "essentially zero" can still be a significant amount; for example, limestone is "insoluble" but tiny amounts do dissolve in water (and form lacy cave formations). Usually, chemists prefer to use the term "slightly soluble." But "insoluble" is still a useful term for things that don't dissolve more than a very tiny amount.

5. Often, when two soluble compounds are mixed a precipitate is formed. Describe, on the atomic/molecular level, what happens in such a reaction.

   When the two soluble compounds meet, they exchange portions that prefer each other's' company to the company of water molecules, and so they fall out of solution. This results in a precipitate.

   If the two compounds are ionic, whenever two ions meet that prefer each other to water molecules, they precipitate as a new ionic compound.

7. What is "soap scum"?

   "Soap" is a sodium-fatty acid salt. "Soap scum" forms when the fatty acid ions interact with calcium ions, forming a calcium-fatty acid precipitate.

Chapter 6

1. "All bonds between atoms involve shared electrons." True or false? Defend your answer.

   Covalent and metallic bonds involve shared electrons. Ionic bonds, in their ideal form, do not involve shared electrons – but absolutely pure ionic bonds don't actually exist.

   Therefore, all bonds between atoms involve electron sharing, even if only a negligible amount.

3. Find at least two pairs of elements on the periodic table that are expected to form an ionic bond.

   Ionic bonds are formed between one metal and one non-metal. Therefore, your pairs should have one metallic element and one non-metallic element.

   Ideally, these elements should be widely separated (otherwise the bond will have significant covalent character) but at this level, "metal + non-metal" is sufficient.

4. Find at least two pairs of elements on the periodic table that are expected to form a covalent bond.

   Covalent bonds form between two non-metallic elements. Therefore your pairs should be of non-metals.

5. Find at least two pairs of elements on the periodic table that are expected to form a metallic bond.

   Metallic bonds form between metal atoms; therefore your pairs should all be metals.

6. Why doesn't hydrogen form ionic bonds as a cation? That is, why isn't *e.g.* hydrogen chloride an ionic compound?

   Hydrogen has only one valence electron. If it formed a cation, that ion would be a naked nucleus – which is not feasible. Therefore, hydrogen can only have an oxidation number of +1 in covalent compounds. (Also, hydrogen is a non-metal; it forms "ionic" compounds with metals as H-minus.)

Chapter 7

2. What is <u>always</u> true of *isomers*?

   Isomers always have the same empirical formula, that is, the same numbers of the same types of atoms. The atoms are simply arranged differently.

3. For each of the chemical reactions shown below, balance the reaction expression.

   $HClO + HCl \rightarrow H_2O + Cl_2$

   Already balanced

   $C_4H_{10} + O_2 \rightarrow CO_2 + H_2O$
   2   13   8   10

   $PbSO_4 + H_2O \rightarrow Pb + PbO_2 + H_2SO_4$
   2   2   1   1   2

   $C_6H_{12}O_6 + O_2 \rightarrow CO_2 + H_2O$
   1   6   6   6

Chapter 8

4. Describe the similarities and differences between a dipole-dipole interaction and a hydrogen bond.

    Dipole-dipole interactions and hydrogen bonds both result from permanent separations of charge, caused by unequal sharing of electrons in covalent bonds. Positively-charged regions on one molecule are attracted to negatively-charged regions on a neighboring molecule.

    Hydrogen bonds are a special, more intense case. A hydrogen atom is involved in a polar bond that results in its having a net, fractional positive charge. Because the hydrogen atom is so small, this charge is concentrated and is strongly attracted to centers of negative charge on neighboring molecules.

5. Describe how London forces are generated. Why are London forces so weak?

    The text describes how London forces are generated. They are weak because they involve very tiny imbalances in electron distribution – like catching a glimpse of someone through a thick fog.

Chapter 9

Calculate the molar masses, also called formula weights, of each of the following compound. The formula is given to you. Don't forget to use the correct unit of mass!

Molar masses are calculated by adding the atomic weights of all the atoms in the compound's formula. That is why they are called formula weights. Two examples:

a. Carbon disulfide, CS_2

One atom of carbon (12.011) and two of sulfur (each 32.06):

12.011 + 2 × 32.06 = 12.011 + 64.12 = 76.131

f. Ammonium phosphate, $(NH_4)_3PO_4$

There are three NH_4 units for a total of three N (14.007) and twelve H (1.008); there are also one P (30.974) and four O (15.999):

3 × 14.007 + 12 × 1.008 + 30.974 + 4 × 15.999 =

42.021 + 12.096 + 30.974 + 63.996 = 149.087

Calculate the number of moles in the given amount of each of the compounds below.

The number of moles in a given amount of a substance is equal to the number of gram-formula weights (that is, formula weights in grams) in that amount. We will use the same two compounds as examples:

20 grams of carbon disulfide, CS_2

The formula weight we calculated above is 76.131 (the units of "formula weight" are grams per mole).

$$20 \; g \; CS_2 \times \frac{1 \; mole \; CS_2}{76.131 \; g \; CS_2} = 0.2627 \; mole \; CS_2$$

500 grams of ammonium phosphate, $(NH_4)_3PO_4$

The formula weight we calculated above is 149.087 g/mol.

$$500 \; g \times \frac{1 \; mol}{149.087 \; g} = 3.3537 \; mol$$

Chapter 10

2. Describe absolute zero, in terms of

    a. how the volume of a gas changes with temperature.

    Absolute zero is the temperature at which the volume of an ideal gas (or the extrapolated volume of a real gas) reaches zero, assuming pressure is held constant.

    b. the kinetic molecular theory of gases.

    Absolute zero is the temperature at which the molecular kinetic energy of a gas is zero. (More correctly, it is the temperature at which **every** molecule of a gas is in the lowest-possible energy state.)

3. Why does a real gas condense to a liquid, or freeze to a solid, before reaching absolute zero?

    An ideal gas is a model in which gas molecules have zero volume, and do not interact with each other at all via intermolecular forces.

    Real gas molecules do not have zero volume, and they have non-zero intermolecular attractions. The closest real gas to an ideal gas is helium: its atoms are very small and "hard," with almost no interatomic attractions. At one atmosphere of pressure, helium liquefies at 4 K, the lowest temperature of any known gas.

Chapter 11

1. Explain why free-radical reactions tend to occur in the gas phase, rather than the liquid or solid phase.

   Free radicals are so reactive that they will react with almost the first molecule they collide with. In a liquid or solid, molecules are in contact with each other and so a free radical will be quenched instantly.

2. Explain why gases are considered *fluids* like water or gasoline.

   Gases flow, like any other fluid.

5. At the beginning of the "For example" section on photosynthesis, it says that for electrons to flow, they need both a source and a place to go. How does this apply to redox reactions?

   Redox reactions require something that is oxidized (gives up electrons) and something that is reduced (gains electrons). The electrons are transferred from the thing being oxidized (the electron source) to the thing being reduced (the electron sink).

6. Is photosynthesis a redox reaction? Discuss how you can tell from the balanced reaction expression:
   $$6\ CO_2 + 6\ H_2O \rightarrow C_6H_{12}O_6 + 6\ O_2$$

   Consider just the oxidation number of carbon. In CO_2, carbon has an oxidation number of +4. In glucose, the six carbons have an oxidation number of zero – the hydrogen and oxygen atoms balance each other perfectly. Since carbon has gone from +4 to zero, it has been reduced. (Oxygen has been oxidized, from −2 to zero.)

Chapter 12

1. Why is ice less dense than liquid water?

    www.edinformatics.com/interactive_molecules/ice.htm shows that in ice, water molecules are arranged in hexagons. Space between molecules is present in the centers of the hexagons. But in liquid water, molecules are in intimate contact with each other. The appearance of intermolecular spaces in ice makes it less dense than liquid water.

2. How does surface tension result from intermolecular forces?

    Molecules in the liquid, clinging to each other, exclude other molecules and even macroscopic objects like hairs and insects. This is analogous to the way that several people's joined hands and arms can support another person.

Chapter 13

2. People sometimes use the fact that you get heat by burning methane – changing it into carbon dioxide and water – to claim that energy is released by breaking chemical bonds. Explain why this is not true, and where the energy release comes from when you burn methane.

    The energy released when methane is burned is a result of the fact that the C-H bonds in methane (and the O=O bonds in oxygen) take less energy to break, than the energy released when C=O and O-H bonds are formed to make carbon dioxide and water.

3. In terms of enthalpy, entropy and free energy, explain what conditions must be true for a process to be *spontaneous*.

    The overall free energy change resulting from the process must be negative. Free energy is a sum of enthalpy and "entropy times −1" – so that if the entropy is negative (a decrease in entropy, which when multiplied by −1 gives a positive number), you can still have a spontaneous process if the enthalpy is negative and large enough to compensate.

7. Charging your cell phone battery always makes the battery warm to the touch. Why is that?

    The heat is a consequence (a manifestation, really) of the fact that entropy increases in any physical or chemical process. Charging the batter is an electrochemical process (see Chapter 18) and the increase in entropy shows up as heat, which is dissipated, increasing the entropy of the battery and its surroundings.

Chapter 14

1. A change from liquid to gas is always endothermic. Why, then, does water evaporate spontaneously?

   Heat (enthalpy) is absorbed by the water that evaporates. There is a large entropy increase involved in spreading that heat widely on the water vapor, enough to overcome the positive enthalpy change.

2. Using the ideas of kinetic molecular theory, explain why melting is endothermic and freezing is exothermic.

   Melting is endothermic because molecules of a solid must absorb energy in order to move faster than they do in the solid phase.

   Freezing is exothermic because molecules of a liquid must lose energy in order to stay in their assigned positions in the solid.

3. Describe what is meant by a *metastable state*. Give an example, and explain why it is metastable.

   A metastable state is one that could (in theory) stay as it is, but is easily perturbed into changing to another, more stable state. An example is a two-by-four balanced on one end: the flat end allows it to stand, but it can be knocked over by a small gust of wind.

7. Why is it that liquids are not found in nature, except under certain conditions? What are those conditions? The diagram immediately above this question may be helpful in answering.

   If you examine the phase diagram in #6, you will see that – below some particular pressure, different for each substance – the substance can be either a gas or a solid, but not a liquid.

   Above some particular pressure and temperature, the substance becomes a supercritical fluid, neither gas nor liquid.

   The liquid state can exist only between those two extremes of pressure. Since most of the universe is at very low pressure, liquids cannot exist in most of the universe.

Chapter 15

2. "In a saturated solution, no more solid is able to dissolve." Explain how and why this statement is false, on the molecular level. In what way is the statement true, on the molecular level?

   On the molecular level, the solute in a saturated solution is in equilibrium between the solid form and the dissolved form. Individual molecules of the solute are constantly dissolving, and constantly being added to the solid state.

   But since these things are happening at the same rate, no more **net** solid dissolves.

5. How can equilibrium be "unbalanced"? What happens to the components of a mixture if their equilibrium is unbalanced?

   If a process normally exists in equilibrium, but the quantities in the process are not at their equilibrium levels, the process is said to be an "unbalanced equilibrium." When an equilibrium process is unbalanced, the forward and backward processes take place at different rates until all components are at their equilibrium levels.

6. In terms of equilibrium, explain why the temperature of boiling water does not rise above 100°C (212°F).

   At the boiling point of water, at the surface water molecules are evaporating just as fast (or a little faster, even) as they are condensing. The water molecules that turn to gas and escape take energy away, while those that are condensing return energy to the liquid, so that there is no net change in temperature.

8. Suppose that we mix equal amounts of water containing natural hydrogen (H_2O or HOH) – natural hydrogen consists almost entirely of the hydrogen-1 isotope – and water containing only the hydrogen-2 isotope, symbolized by "D" (D_2O or DOD). After a few days, all the water molecules in the mixture will have one atom of hydrogen-1 and one atom of hydrogen-2 (DOH). Explain this. What could be going on to make this happen?

   The easiest way to explain this is to suppose that there is some process by which hydrogen atoms are continuously exchanged between water molecules. In fact, we know of such a process: an acid-base reaction in which water is both the acid and the base.

Chapter 16

2. Explain osmosis in terms of entropy.

    In osmosis, a solute is diluted. This spreads the solute more thinly and widely, and increases the overall entropy – making the process favorable.

4. Explain freezing-point depression in terms of entropy.

    At the freezing point, the liquid is changing to a solid. But if the liquid has a solute dissolved in it, it is likely that the solid will not dissolve that solute nearly as well, perhaps not at all. Thus, freezing means that entropy will decrease as the solvent and solute are separated and the remaining liquid becomes a more concentrated solution.

    However, reducing the temperature reduces the effect of this unfavorable entropy change, and so we see the freezing point lowered.

5. Why must energy be expended to get reverse osmosis to happen?

    a. Explain using concepts on the molecular scale.

    In osmosis, water molecules flow more rapidly from the pure-water side to the solution side of a membrane. To get that flow to reverse, energy must be applied to the solution side to make the water molecules on that side move faster or more forcefully, so that they can flow back to the pure-water side of the membrane more rapidly.

    b. Explain in terms of entropy.

    In osmosis, water flows through a membrane in such a way as to dilute a solute, increasing the overall entropy. To reverse this – to cause a decrease in overall entropy – energy must be used. (Any process that decreases entropy requires an energy input of some sort; this energy input results in an increase in entropy somewhere else.)

Chapter 17

2. Explain how a catalyst works, in terms of activation energy.

   A catalyst lowers the effective activation energy of a reaction. It does this by **providing a different reaction pathway** that has a lower activation energy than the pathway that would be used in the absence of the catalyst.

4. Explain why decreasing the temperature at which a reaction happens will make that reaction happen more slowly.

   Decreasing the temperature means that molecular energies are lower. At lower energies, collisions will be neither as frequent nor as energetic, resulting in a lower number of reaction collisions per unit time.

6. Explain why, if a reaction involves a reduction in entropy, increasing the reaction temperature will slow the reaction.

   The influence of entropy is increased with increasing temperature: $\Delta G = \Delta H - T\Delta S$, where T is temperature and S is entropy.

   A decrease in entropy is unfavorable and will decrease the reaction rate; increasing the temperature increases the influence of entropy on the reaction, making it even slower.

Chapter 18

1. A chemical light stick and the fading of colors in the sun are both examples of photochemistry. Describe the similarities and differences between the two processes **as photochemical reactions**.

   A chemical light stick uses a chemical reaction to produce light. When colors fade in the sun, sunlight is causing a chemical reaction. In both cases, light and matter are interacting: in one, a chemical change produces light, and in the other, light produces a chemical change.

3. What three components are necessary for a battery?

   Batteries require a cathode, an anode and an electrolyte:

   A cathode is an oxidizing agent, and absorbs electrons from the anode (a reducing agent). The electrolyte keeps the cathode and the anode from coming into electrical contact inside the battery, and also supplies ions to compensate for the unbalanced charge that builds up as negative electrons are moved from anode to cathode.

5. What is the essential difference between a primary battery and a secondary battery? Which is a true energy source?

   A battery is a device in which chemistry and electricity interact: when a battery charges, electricity is used to cause a chemical reaction; when it discharges, a chemical reaction is used to produce electricity.

   A primary battery is one that cannot be recharged; this means that it can move only one way: a chemical reaction acts as the source of electricity.

   A secondary battery is one that can be recharged, so that its chemistry can move in both directions. Thus, it functions as a device for storing electricity.

6. Describe what a sacrificial anode does.

   A sacrificial anode donates electrons to a metal object that would otherwise corrode (be oxidized) by its surroundings. By so doing, the sacrificial anode is itself corroded ("sacrificed").

www.ingramcontent.com/pod-product-compliance
Lightning Source LLC
Chambersburg PA
CBHW040820180526
45159CB00001B/6